A Symphony of Light: Unveiling the Hidden Dance of Molecules with Lasers

Sana

Table of Contents

Chapters

Chapter 1
INTRODUCTION

1.1 Ultrafast Science

Since the discovery of the laser [1, 2] and the subsequent demonstration of nonlinear optics
[3, 4], one of the areas that has seen a great amount of interest is the use of lasers to
study the dynamics of various systems. This is generally done by taking advantage of the
ability to create a pulsed laser [5]. This occurs when the spectral phase and amplitude of
a laser coherently combines to create an intense burst of light as a function of time. As an
example, a typical approximation is to assume that the electric field $\mathcal{E}(t)$ of a laser pulse
can be written as a Gaussian, such as

$$\mathcal{E}(t) = \mathcal{E}_0 e^{-2\ln 2(\frac{t}{\Delta t})^2} \cos(\omega_0 t + \theta(t)) \tag{1.1}$$

where Δt is the pulse duration, ω_0 is the carrier frequency, and $\theta(t)$ determines the temporal
relationship between the frequency components that are within the pulse bandwidth. An
example of just such a pulse shape is shown in figure 1.1 in both time and frequency, and it
can be shown that the pulse duration Δt is inversely proportional to the spectral bandwidth
$\Delta\omega$. This entails that larger bandwidths are required to achieve shorter Fourier-transform-
limited pulse durations.

The evolution of pulse duration since the 1960s is highlighted in figure 1.2, and it can
be seen that there has been a tremendous push by the community to produce ever shorter
pulses. The need for ever shorter pulse durations is rooted in how dynamics can be extracted from a system using pulsed lasers, and this is typically done through a pump/probe
experiment. A schematic of a pump/probe experiment is shown in figure 1.3. The core
concept revolves around the use of one laser pulse to induce a change within a physical
system and a second laser pulse to probe the perturbed system at a variable time later,
known as the time delay τ. Each of these time delays can be thought of as a single film
frame, and by varying the time delay, one can construct a movie of the dynamics induced
by the pump pulse.

nature photonics
FOCUS | REVIEW ARTICLES
PUBLISHED ONLINE: 28 FEBRUARY 2014 | DOI: 10.1038/NPHOTON.2014.28

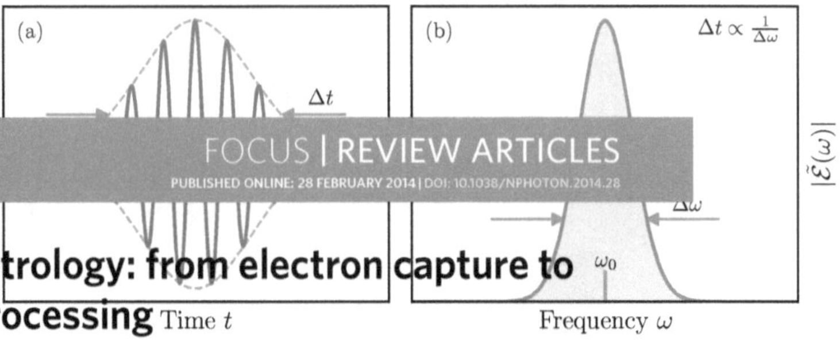

Figure 1.1: Example of a typical Gaussian pulse in time (a) and frequency (b). The pulse [illegible] the relationship between them

Attosecond metrology: from electron capture to future signal processing

Ferenc Krausz[1,2] and Mark I. Stockman[1,2,3]

The accurate measurement of time lies at the heart of experimental science, and is relevant to everyday life. Extending chronoscopy to ever shorter timescales has been the key to gaining real-time insights into microscopic phenomena, ranging from vital biological processes to the dynamics underlying high technologies. The generation of isolated attosecond pulses in 2001 allowed the fastest of all motions outside the nucleus — electron dynamics in atomic systems — to be captured. Attosecond metrology has provided access to several hitherto immeasurably fast electron phenomena in atoms, molecules and solids. The fundamental importance of electron processes for the physical and life sciences, technology and medicine has rendered the young field of attosecond science one of the most dynamically expanding research fields of the new millennium. Here, we review the basic concepts underlying attosecond measurement and control techniques. Among their many potential applications, we focus on the exploration of the fundamental speed limit of electronic signal processing. This endeavour relies on ultimate-speed electron metrology, as provided by attosecond technology.

Attosecond science represents the culmination of the long-standing investigation of fast-evolving phenomena in nature. After the concepts of pump–probe spectroscopy[1] and optical synchronization[2] were introduced in the nineteenth century, subsequent progress in improving the temporal resolution of time-resolved measurements was dictated by the development of light sources capable of producing ever-shorter pulses and of techniques for their measurement. Several technological revolutions were required to advance ultrafast science from the nanosecond scale, which can be accessed by incoherent light sources, to the femtosecond regime, and eventually to the attosecond regime. The laser[3,4] (for which Basov, Prokhorov and Townes were awarded the Nobel Prize in Physics in 1964) provided the basis for all these advances, thanks to its ability to emit perfectly sinusoidal light waves over many seconds, which consist of quadrillions (10^{15}) of identical wave cycles. This coherence opened the door for measuring time over long periods and brief intervals with unprecedented precision and resolution, respectively.

Perturbative nonlinear optical effects[5,6] (for which Bloembergen was awarded the Nobel Prize in Physics in 1981) allow the optical properties of materials, such as their refractive index and absorptivity, to be varied as a function of the cycle-averaged intensity of intense laser radiation. Some of these changes can be harnessed to create light modulators, which alter the phase and/or amplitude of transmitted or reflected laser light in proportion to its intensity. Integrating such a device in a broadband laser oscillator permits the eigenmodes of the laser cavity modes to be phase locked. A second revolution, namely in nonlinear optics[7], thus led to the generation of femtosecond laser pulses[8]. By the turn of the millennium and some 15 years after the seminal report on three-cycle optical pulses by Shank and co-workers[9], the evolution of femtosecond technology approached its ultimate limit — 1 the wave cycle[10]; this was achieved through realizing broadband dispersion control using chirped multilayer mirrors[11] (Fig. 1). The envelope of the fastest, few-femtosecond laser pulses is short enough to temporally resolve any molecular dynamics[12] (for which Zewail was awarded the Nobel Prize in Chemistry in 1999), but it is insufficient to capture atomic-scale electronic motion.

Strong-field interactions driven by few-cycle light pulses[10] enabled the electronic responses of matter to be controlled and

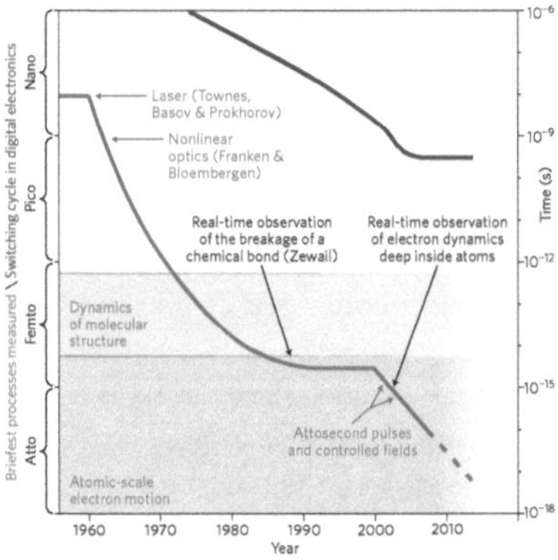

Figure 1 | Evolution of ultrafast science and digital electronics. Briefest [illegible] (inverse of the clock rate) of digital processors (upper blue line). Discontinuities in the derivative of the briefest measured interval against [illegible] coherent light waves generated by lasers and nonlinear optical effects were used to produce ultrashort pulses from these waves. At the beginning of the new millennium, attosecond pulse generation and measurement[18] were realized, and controlled light fields for precision attosecond control[28] and metrology[34] were generated.

Figure 1.2: Pulse duration as a function of time since the discovery of the laser in 1960. A femtosecond barrier was reached in the 1980s as laser pulses approached a pulse duration of only a single period of the carrier frequency. High-harmonic generation allowed for pulses to push beyond this barrier. Adapted from [1].

measured within an optical cycle, that is, on an attosecond timescale. Just as cycle-averaged-intensity-driven perturbative nonlinear optics had led to femtosecond pulses, nonperturbative

[1]Fakultät für Physik, Ludwig-Maximilians-Universität, Am Coulombwall 1, D-85748 Garching, Germany, [2]Max-Planck-Institut für Quantenoptik, Hans-Kopfermann-Strasse 1, D-85748 Garching, Germany, [3]Department of Physics and Astronomy, Georgia State University, Atlanta, Georgia 30303, USA. e-mail: krausz@lmu.de; mstockman@gsu.edu

The temporal resolution of this "movie" is ultimately limited by the pulse duration of both the pump and probe because time is not an observable in these experiments and it can only be deduced by varying the delay. Thus, to access processes that happen on a short timescale you need a pulse with an even shorter pulse duration. The types of processes that are unlocked as the pulse duration get shorter and shorter is shown in figure 1.2. Once the pulses are in the femtosecond (10^{-15} s) regime, then the dynamics of molecular structure and motion can be observed in real-time, and the experimental demonstration of this was awarded the Nobel Prize in 1999 [7]. To go beyond this regime to study the dynamics of electrons on their natural time scale, then one needs to have attosecond (10^{-18} s) pulses. The breakthrough that enabled such pulses to be generate is high-harmonic generation, and it will be discussed in the next section.

1.2 High-Harmonic Generation

High-harmonic generation (HHG) is a strong-field process that occurs during the interaction of an intense, femtosecond pulse with a gas[1] medium. The strong-field regime that needs to be reached in order for this process to occur is characterized by the strength of the laser field being comparable to the binding energy of the atom, and this is generally when the intensity is above 10^{13} W/cm^2. The first demonstration of high-harmonic generation occurred in late 1980s [9, 10], and the spectrum of light that was observed after interaction of an intense femtosecond pulse with rare gas atoms consisted of a plateau of odd harmonics (odd integer multiples of the fundamental photon energy) well into the extreme ultraviolet (XUV) photon energies. The high harmonic order of these processes and the plateau that they form were a clear signal that the physical process behind their generation was non-perturbative [11].

The mechanism behind this was not explained until 1993 by Schafer *et al.* [12] in terms of an electron recollision process. Shortly thereafter, Corkum's paper [13] broke down this recollision process in terms of a simple three-step model consisting of ionization, propagation, and recombination. The three steps in this model are shown schematically in figure 1.4. The first step in the three-step model is tunnel ionization where the strong laser field suppresses the atomic potential enough to allow the electron to tunnel through the potential energy barrier. Once freed, the electron then undergoes the second step in the three-step process which is propagation in the field. During this propagation, the electron picks up kinetic energy from the field and is driven back to the parent ion. The maximum kinetic energy that can be gained from this process is approximately $3.17U_p$, where the

---

[1]High-harmonic generation has also been observed in condensed matter as well [8]. However, high-harmonics generated from solids will not be used in this book. So, for the sake of clarity and brevity, those details will be left out of this discussion and only gas phase HHG will be discussed.

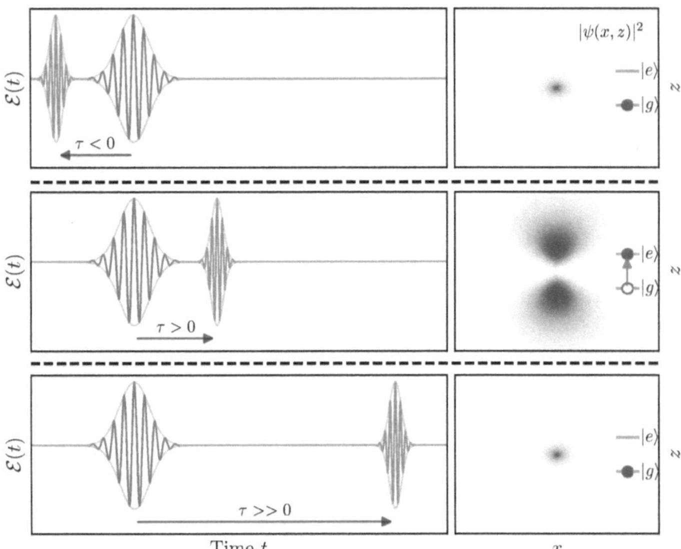

Figure 1.3: Schematic of a pump/probe experiment as a series of frames constituting a movie. Each frame is given by a different time delay between the pump (red pulse) and the probe (blue pulse). In this simple example an electron in a hydrogenic atom is excited from the ground state $|g\rangle$ to an excited state $|e\rangle$ that has some finite lifetime. The state of the atom that the probe pulse observes is shown in the panels of the right. If the probe pulse arrives before the pump, then the ground state is observed. If the probe pulse arrives after excitation by the pump pulse and within the lifetime of the excited state, then the probe pulse observes the excited state. If the probe pulse arrives much later than the pump pulse (beyond the lifetime of the excited state), then the electron has relaxed and only the ground states is observed.

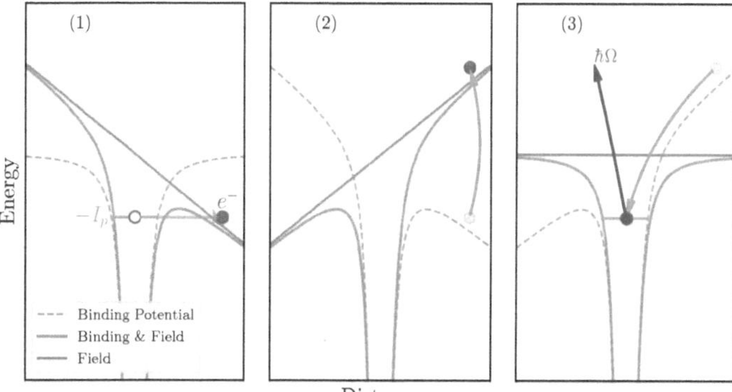

Figure 1.4: Recollision model of high harmonic generation. The three steps are (1) tunnel ionization, (2) propagation/acceleration in the laser and Coulomb fields, and (3) recombination and photoemission.

ponderomotive energy U_p is given by

$$U_p = \frac{e^2 \mathcal{E}_0^2}{4m\omega_0^2} \propto I_0 \lambda^2 \tag{1.2}$$

for a laser of wavelength λ and peak intensity I_0. Once the electron is driven back to the parent ion there are a few possible interaction pathways consisting of inelastic scattering, elastic scattering, and recombination. The pathway of interest is recombination, and when this occurs the excess energy picked up by the electron from the laser field is released in form of a photon of energy $\hbar\Omega$. This means that the maximum photon energy that can be generated in this classical process (also known as the cutoff) is given by the relationship

$$\hbar\Omega_{\text{cutoff}} = I_p + 3.17 U_p, \tag{1.3}$$

where I_p is the ionization potential of the neutral atom.

This three-step process repeats every half-cycle of the laser field, and as a consequence of this, the temporal structure of the generated XUV light is that of a "train" of short pulses every half-cycle of the fundamental field, as is shown in figure 1.5. This structure is referred to as an attosecond pulse train (APT), and it was first experimentally measured in 2001 [14], showing that each burst in the APT has a pulse duration on the order of a few hundred attoseconds. It is possible to isolate a single attosecond burst to create an isolated attosecond pulse (IAP), however this generally involves a single-cycle pulse [15] or

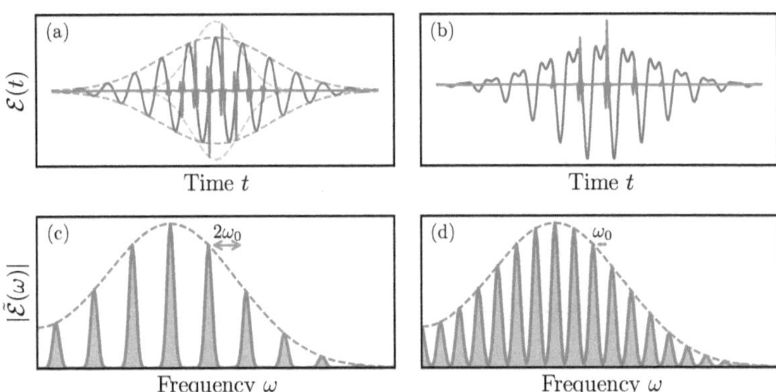

Figure 1.5: (a) Example of the electric field of an XUV APT. There is a burst every half-cycle of the fundamental field period τ_0. (b) Spectral amplitude of APT. Harmonics are separated by $2\omega_0 = 4\pi/\tau_0$. Bandwidth of each harmonic is determined by the number of cycles in the fundamental pulse, and the overall bandwidth is determined by the pulse duration of each burst in the train. (b) Example of a two-color $\omega - 2\omega$ fundamental field. The asymmetry of the pulse means there is a burst once every cycle. (d) The spectral amplitude for the asymmetric field, and now there are even and odd harmonics. The dashed line in (c) and (d) represent the spectral amplitude of just one of the pulses in the pulse train.

an engineered time-dependent ellipticity across the pulse duration [16]. These methods are
not used in the experiments described herein, however there is another method that can
modify the generated harmonic spectrum that will be used. This method involves using a
nonlinear crystal to generate the second harmonic of the fundamental, and when this pulse is
temporally overlapped with the fundamental pulse it creates an asymmetric field, as shown
in figure 1.5. This asymmetric field means that there is now an attosecond burst only once
per cycle of the fundamental. The effect that this has on the spectrum is that there are
now both even and odd harmonics. When the bandwidth of each individual harmonic is
large enough, then the harmonic comb can be treated like a pseudo-continuum of photon
energies. This idea of two-color generation can be extended to pulses with incommensurate
frequencies to generate an IAP or harmonics of the beat frequency [17, 18].

1.3 Transient Absorption Spectroscopy

Now that the basics of high-harmonic generation has been laid out, the question becomes:
What types of experiments can be performed with this unique light source? Generally
speaking, there are two main classes of experiments that can be conducted. One involves
using the XUV APT to ionize a gas and collect the energy-resolved photoelectron spectrum,
and the other involves measurement of the harmonic spectrum using a photon spectrometer.
Collecting photoelectrons allows for measurement such as RABBITT [14] and streaking [15]
when an IR field is used to dress the gas that is being ionized by the APT. These techniques
will not be used in any of the experiments described herein. Instead, the harmonic spectrum
will be measured using a photon spectrometer. This gives access to the spectral amplitude
of the XUV APT, and this enables transient absorption spectroscopy to be performed.

A schematic of how transient absorption spectroscopy works in gases is shown in figure
1.6. This is fundamentally a pump/probe measurement where an XUV APT serves as the
probe and an IR pulses acts as the pump. The idea is that the XUV pulse induces a polarization within the gas which decays on a timescale that is set by the lifetime of the excited state
that the XUV populates. When this time-domain picture is Fourier transformed through
the measurement of the spectral amplitude, the absorption spectrum of the gas medium
can be deduced. This becomes transient absorption when an IR pulse is introduced at a
variable time delay. This IR pulse will perturb the induced polarization, and the resulting
absorption spectrum will be modified. This modification of the absorption spectrum as
a function of time delay can be used to study the dynamics of the excited states within
the gas. This measurement is at the core of the work that was done in this book,
especially Chapters 5 and 6, and more explicit details about transient absorption will be
discussed in those Chapters.

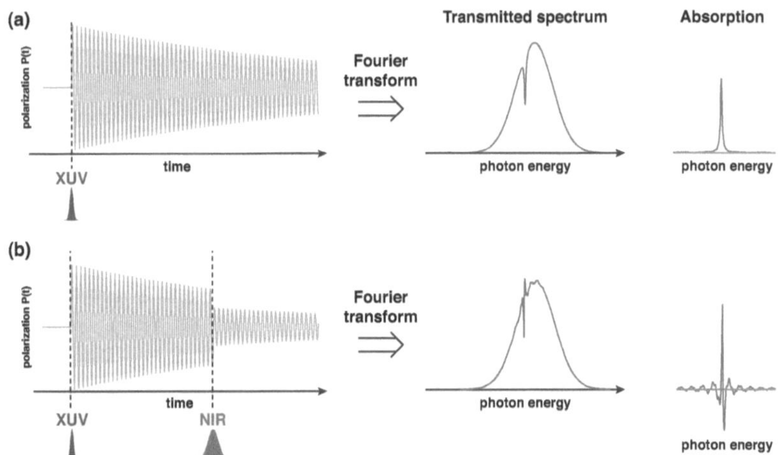

Figure 1. Simplified diagram of transient absorption. (a) XUV-only absorption measurement. The XUV pulse induces a polarization in the sample (left). The transmitted spectrum (middle) shows a linear absorption lineshape and the calculated absorption (right) displays a single peak. (b) The NIR pulse then perturbs the time-domain polarization at some time delay (left), altering the transmitted spectrum (middle) and creating sidebands or altered absorption lineshapes in the expected absorption (right).

femtoseconds for a short-lived autoionizing state to nanoseconds for an atomic Rydberg level. The subsequent NIR pulse perturbs this polarization when it interacts with the sample. The macroscopic polarization is formed by a collection of oscillating dipole moments, which emit light. This emission is then dispersed in a spectrometer, essentially measuring the Fourier transform of the time-dependent signal. Absorption features are observed due to destructive interference between the emission from the excited sample and the transmitted broadband pulse. Because the induced polarization has been perturbed by the NIR pulse, the measured absorption features change. This is shown schematically in Figure 1.

The temporal resolution of a transient absorption experiment depends on the duration of the laser pulses used in the experiment while the resolution in the frequency domain depends only on the energy resolution of the spectrometer, as described by Pollard and Mathies [5]. This is because the final absorption measurement, which is the signal at the detector integrated over all times, is not time-dependent. Therefore, attosecond transient absorption measurements can be performed with both excellent temporal resolution and good spectral resolution.

In conventional transient absorption spectroscopy, a weak broadband pulse is used to measure the absorption spectrum of a sample. A pump pulse initiates dynamics, and changes in absorption features are observed when the broadband pulse follows the pump pulse. These methods have been successfully extended to absorption experiments, the range of time delays when the pulse follows the XUV pulse might be deemed uninteresting. No dynamics can occur before the pump pulse has interacted with the system. However, important recent results in attosecond science have capitalized on transient absorption experiments in which the NIR pulse overlaps with or follows after the attosecond pulse. This Frontiers article focuses on the rich variety of dynamics that can be observed and characterized in these two regions of NIR/attosecond pulse time delay.

2. Experimental techniques of attosecond transient absorption

A generalized experimental apparatus for attosecond transient absorption is shown in Figure 2. Similar apparatuses used by other groups are described in detail by Chini et al. [9] and Ott et al. A NIR pulse (duration of approximately 7 fs) is first split by a beam splitter in an interferometer. One arm of the interferometer generates the attosecond pulse, while the other arm acts as the NIR pulse. The XUV pulse generating arm may pass through optics to implement an attosecond optical gating technique, such as Double Optical Gating [11], which manipulates the laser field in time to produce a single attosecond pulse. The pulse then enters

Chapter 2
TRANSIENT ABSORPTION BEAMLINE

2.1 Introduction

In order to perform the experiments described in this book, it was necessary to con-struct a purpose-built experimental apparatus, and this is what will be referred to as the Transient Absorption Beamline (TABLe). The TABLe was designed, constructed, and commissioned with the express intent to perform attosecond transient absorption experiments in the XUV energy range using a HHG source. The TABLe was designed, constructed, and commissioned by my fellow graduate student, Greg Smith, and myself. Each of us took responsibility for certain design aspects. My contributions to the design and capabilities of this beamline is discussed within this Chapter, and further details can be found in Greg Smith's book [20].

2.2 Laser System

To begin, it is important to describe the laser system that was used for all of the experiments described in this book. This laser system in question is a commercial system from Spectra-Physics called the Spitfire Ace PA. It is a titanium-doped sapphire (Ti:Sapph) laser that is based on chirped-pulse amplification (CPA).[2] A schematic of this CPA system is shown in figure 2.1. The first main component is the oscillator, and this is a laser cavity that is designed to produce pulses via mode-locking. In the Spitfire, this oscillator is acousto-optic mode-locked, and it produces pulses of 10 nJ at repetition rate of 80 MHz with a bandwidth of 25 nm centered at 800 nm. The next component is a stretcher, which is used to reduce the peak intensity of each pulse before amplification to avoid damaging optics. Following the stretcher, the next component is the regenerative amplifier (regen). The regen is a cavity-based amplifier that uses a Pockels cell as a pulse picker to allow a single oscillator pulse into the cavity to be amplified. The pulse is kept in the cavity for

---

[2] The development of the CPA laser system by Gérard Mourou and Donna Strickland was awarded the Nobel Prize in 2018.

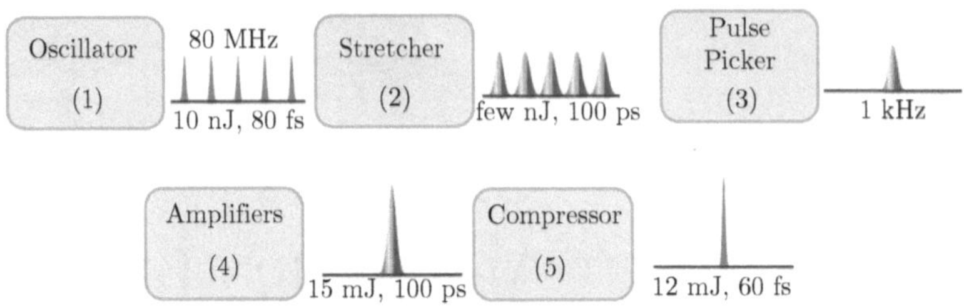

Figure 2.1: Schematic of a CPA laser system. Adapted from [21].

a set number of round trips before it is released, and at this point the pulse is now at the mJ level of pulse energy. Following this stage of amplification, another stage is required to further amplify the pulse energy, and this is accomplished using single-pass amplifier. For the Spitfire system, the energy after these amplification stages is 15 mJ at a repetition rate of 1 kHz, however the pulse is still stretched in time. So, the last component is the compressor, and this uses a diffraction grating to compress the pulse down to the final pulse duration. After all of these stage, the Spitfire is capable of outputting pulses with a pulse duration of 60 fs and a pulse energy of 12 mJ with a repetition rate of 1 kHz.

Many experiments can be done using these 800 nm pulses, however for the experiments described herein, wavelengths longer than 800 nm are required. To accomplish this, an optical parametric amplifier (OPA) is used to convert the 800 nm pulses into longer wavelengths, and this allows for wavelength tunability. The basic idea of the OPA is to use nonlinear media to convert one larger photon (800 nm pump) into two smaller photons (1200-1600 nm signal and 1600-2400 nm idler) [11]. The OPA used with the Spitfire is a commercial HE-TOPAS Prime from Light-Conversion. The basic optical layout is shown in figure 2.2, and consists of an initial white-light continuum generation followed by a series of nonlinear Beta Barium Borate (BBO) crystals. These BBO crystals are optimized for difference frequency generation to amplify the 1200-1600 nm signal. After these stages of amplification, the TOPAS can output a combine energy of 6 mJ for the combined signal and idler beams. The pulse duration of these pulse can vary depending upon alignment and the wavelength selected, however it is generally around 60 fs. An example of a typical pulse in both intensity and spectrum is shown in figure 2.3. This was measured using a standard second-harmonic generation FROG setup which allows for full characterization of the pulse by extracting both the spectral amplitude and phase [22].

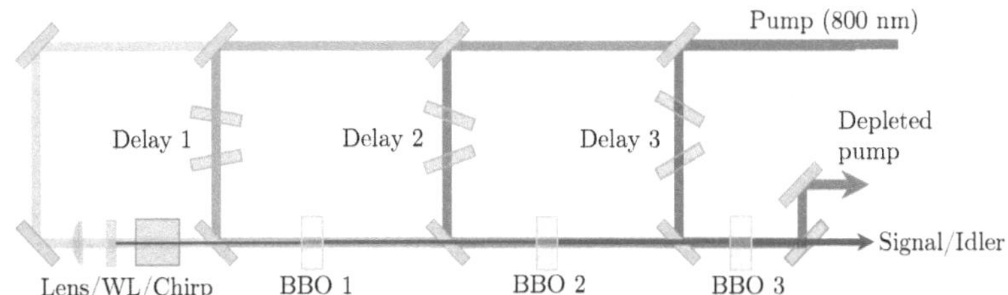

Figure 2.2: Schematic of TOPAS OPA. The pump beam enters with a pulse energy of 12 mJ before being split three times. The lowest pulse energy beam is used to generate white light that is used to seed difference frequency generation in the successive BBO crystals.

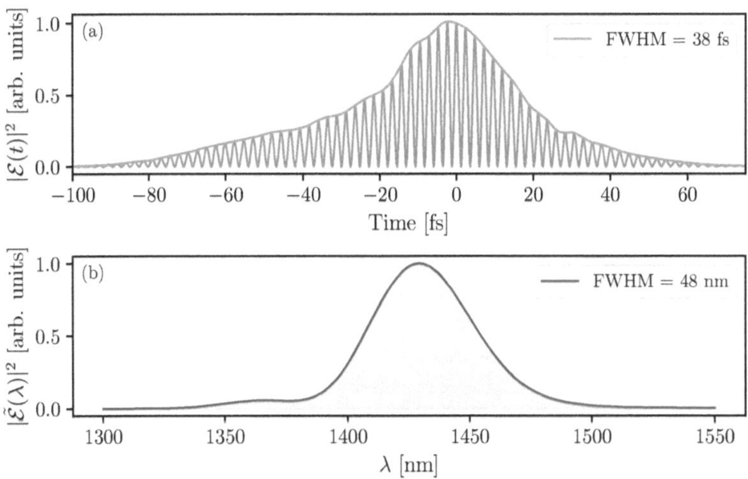

Figure 2.3: (a) Intensity profile of TOPAS output for a signal wavelength of 1430 nm. (b) Spectrum of TOPAS output for the same wavelength.

Figure 2.4: Render of the Transient Absorption Beamline (TABLe). The main vacuum chambers are highlighted.

2.3 Beamline Design

To begin discussion of the design of the TABLe, it is important to lay out the requirements that needed to be met when designing the apparatus. The primary feature common to all of the experiments that use TABLe is that they will involve XUV pulses that have bandwidth between 26 and 300 eV. One of the challenges of working with wavelengths in this XUV region is that they require the use of high vacuum systems to perform most experiments. The fundamental reason for this is due the fact that these wavelengths are ionizing radiation, and they will be quickly absorbed in air. For example, at 50 eV the attenuation length[3] is approximately 50 μm for N_2 at 760 Torr [23]. Therefore, experiments must be performed in vacuum chambers generally in the high vacuum regime below 10^{-6} Torr. An additional requirement is that these experiments are generally pump/probe experiments where the XUV pulse serves as either the pump or the probe and the other role is fulfilled by an IR pulse. Thus, an interferometer needs to be designed where an input IR beam is split into two arms, one of which will have XUV generated in it. A final requirement is that the beamline needs to be able to accommodate different end stations. This is a critical feature because it greatly expands the capabilities of the apparatus beyond the initial experiments that the TABLe was commissioned for, namely transient absorption experiments.

---

[3]Attenuation length is given by the length that the intensity fall to $1/e$ as the beam propagates through a medium.

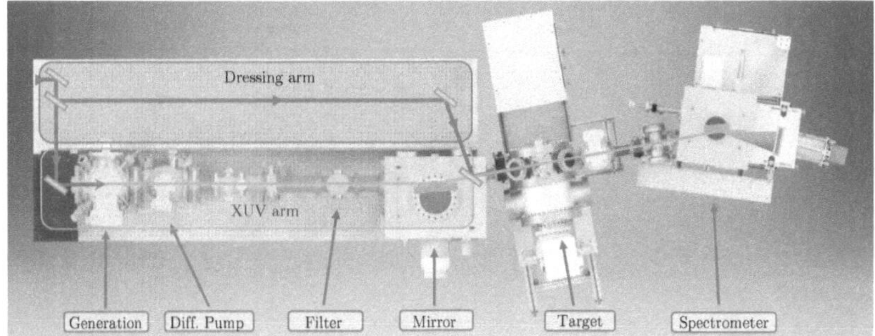

Figure 2.5: Top-view render of the Transient Absorption Beamline (TABLe). The main vacuum chambers are highlighted, as well as the generation layout of the interferometer with the dressing and XUV arms. Further details of the exact optical layout can be found in figure 2.17.

The final design of the TABLe is shown in figure 2.4, and a top-view of the beamline with a rough schematic of the interferometer is shown in figure 2.5. There are many details about the design and construction that will be left out of this book, however they can be found in great abundance in Greg Smith's book [20]. With that being said, the general structure of the vacuum system will be described first, and this consists of several main sections. The first being the generation chamber which contains the gas source used for HHG. This chamber is capable of accommodating a variety of gas sources consisting of continuous gas jets, pulsed gas jets, low pressure gas cells, and high pressure gas cells. Each experiment has different requirements and focal geometries that make any one of these options more advantageous than the others, however the two options that will be used in this book are the high pressure cell and the pulsed gas jet. The details of the high pressure cell can be found in [20]. A picture of the high pressure cell with a laser filament going through it is shown in figure 2.6. The distinguishing feature of the high pressure cell is its ability to achieve interaction pressures up to several bar through the use of a pair concentric tubes that the allow for differential pumping. The primary disadvantages are the difficulty in initial alignment and the restrictive set of four holes that the laser must propagate through. This makes the high pressure cell untenable for some experiments, such as those described in Chapters 3, 4, and 6. An alternative to the gas cell is the pulsed gas jet. This works by using a piezoelectric actuator to block gas from exiting a gas nozzle and only let out a pulse of gas at a fixed frequency and duty cycle. This operation mode means that the average background pressure within the vacuum is lower than just a simple continuous gas jet operation, but the interaction pressure can be the same. A picture of

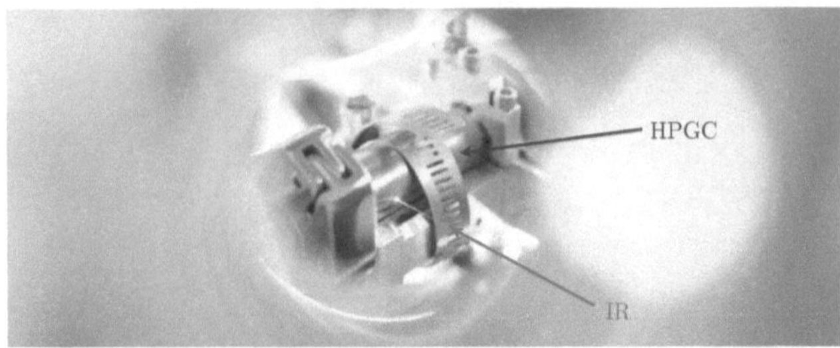

Figure 2.6: Camera image of IR laser filament pass through the high-pressure gas cell (HPGC) during alignment at atmospheric conditions.

the pulsed jet in operation is shown in figure 3.17.

Following a differential pumping section, there is a metallic spectral filter assembly. This is used to place a metallic filter that is generally on the order of a few hundred nanometers into the beam path of the XUV. The primary reason for this is to block the co-propagating IR field that is used to generate the XUV pulse. However, it can also be used to shape the spectrum and phase of the XUV pulse to suit the experimental requirements [24]. The transmission function of three of the most common filter materials is shown in figure 2.7. As can be clearly seen, different materials have significantly different bandpass windows, and this is due to each elements atomic structure and the transition energy of the core level. For example, Al is the filter that is most commonly used for these experiments, and it transmits reasonably well below the L-edge which occurs at 72.3 eV, however it strongly absorbs above this energy.

The next main section of the TABLe that must be described is the mirror chamber. This section houses the primary XUV focusing optic, the ellipsoidal mirror. Again, much greater detail can be found in [20], however the key features of this special optic will be discussed here. A significant challenge with working in this energy range is the lack of efficient and broadband optics. This is due to the strong absorption by most materials when they are used near normal incidence. There are two main methods that have been employed to overcome this: grazing incidence optics or multilayer optics [25]. Each have their advantages and disadvantages, but generally grazing incidence optics operate over a much larger bandwidth. This was a key design criteria for the TABLe, so a grazing incidence optic was selected to refocus the XUV beam. There are several designs that can be used to refocus a beam using a grazing incidence optics, and the one that was selected was the ellipsoidal mirror. The significant advantage of this type of optic is that it can re-image a point source aberration free with demagnification. The geometric optics principle behind

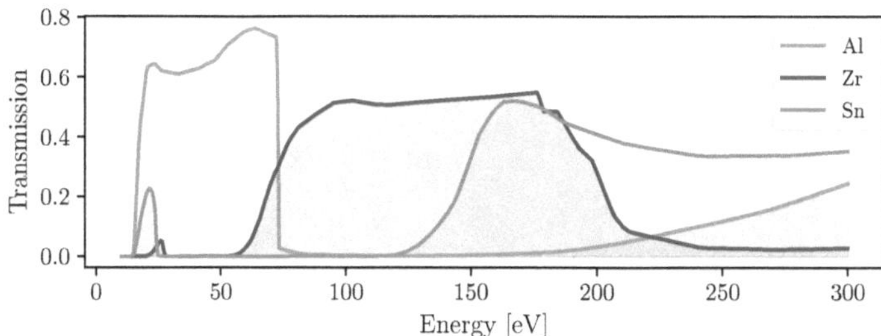

Figure 2.7: Transmission as a function of photon energy for several metallic filters. All filters are 200 nm thick.

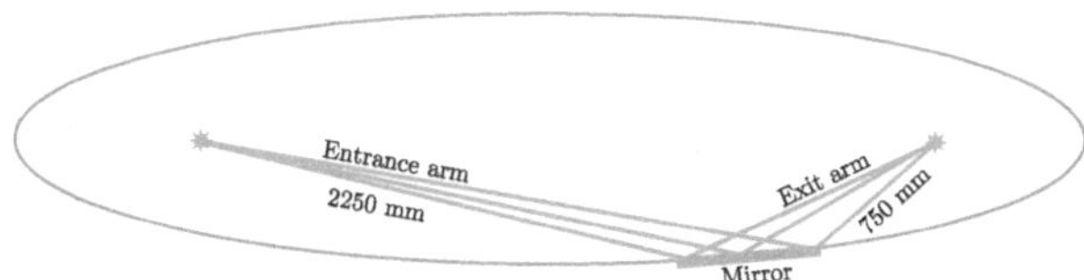

Figure 2.8: Schematic of ellipsoidal mirror. From geometric optics, the ellipsoidal mirror will focus all rays from a point source at one foci of the ellipse to the other foci. The ratio of the exit arm to the entrance arm is the magnification provided by the mirror.

this is highlighted in figure 2.8. The ellipsoid that was chosen for the TABLe has an entrance arm of 2250 mm and an exit arm of 750 mm, thus giving a demagnification of 3. This allows for smaller XUV spot sizes to be used.

The next major section of the TABLe is the target chamber. As the name implies, this is where the sample of interest for an experiment is placed. This chamber is capable of housing both condensed matter samples in the form of free standing membranes and a gas cell to perform gas phase measurements. The two configurations that the target chamber can operate in is shown in figure 2.9.

Finally, the last major component of the TABLe that needs to be discussed is the photon spectrometer. This home-built spectrometer was commissioned to perform the experiments described in this book as well as those described in [20]. The design criteria that this spectrometer had to meet was that it had to be a high resolution spectrometer covering a spectral range from 20 eV all the way through the Carbon K-edge at 284 eV. This was chosen because of eventual plans to perform transient absorption experiments in the water-

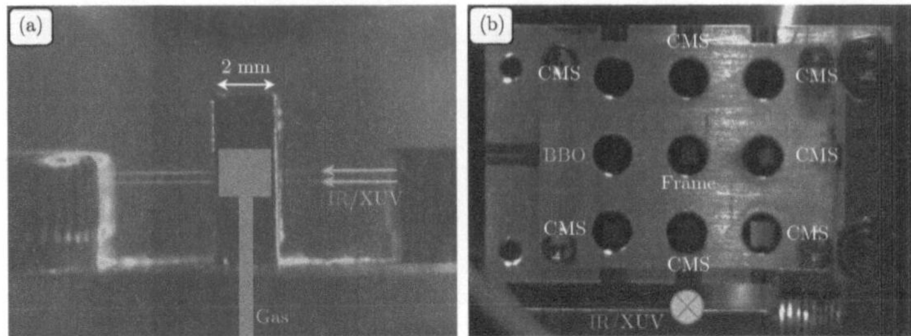

Figure 2.9: (a) Camera image of the gas cell used for gas phase experiments. Two sources IR sources are generating a filament. Image was taken while chamber was vented and at ambient pressure. (b) Camera image of the condensed matter sample holder. This was taken *in-situ* while under vacuum. The sample holder is generally configured to carry six condensed matter samples (CMS), one BBO to find temporal overlap, and one empty frame to perform knife-edge measurements. The IR/XUV beams propagate into the page for this image.

window at the Carbon K-edge, while still maintaining the ability to do more reasonable experiments at lower photon energies in the 20 - 100 eV range.

This is a massive bandwidth to cover with high resolution, so the decision was made to use two different gratings to cover this full spectral range. The gratings that were selected are concave spherical variable line spaced (VLS) gratings produced by Hitachi. A schematic of the grating is shown in figure 2.10, and the relevant design parameters for the gratings are given in table 2.1. The advantage of a VLS grating design is that they combine the focusing and dispersive optics that are generally needed into a single optic that performs both functions. Furthermore, these gratings are designed to focus a specific spectral bandwidth onto a plane (as opposed to the typical Rowland circle [26, 27]), and this means that a large bandwidth can be in focus on a detector at a single position. This can be seen in figure 2.11 where the focal plane for a range of wavelengths is shown for both gratings. This is calculated using the relationship between the entrance r and exit arm r' at an incidence angle θ_i is given by

$$r' = \frac{rR(\cos\theta_r)^2}{m\lambda\sigma_0 b_2 + r(\cos\theta_i + \cos\theta_r) - R(\cos\theta_i)^2} \tag{2.1}$$

where R is the radius of curvature, m is the diffraction order, θ_r is given by

$$\sin\theta_i + \sin\theta_r = m\sigma\lambda, \tag{2.2}$$

Grazing-Incidence Type

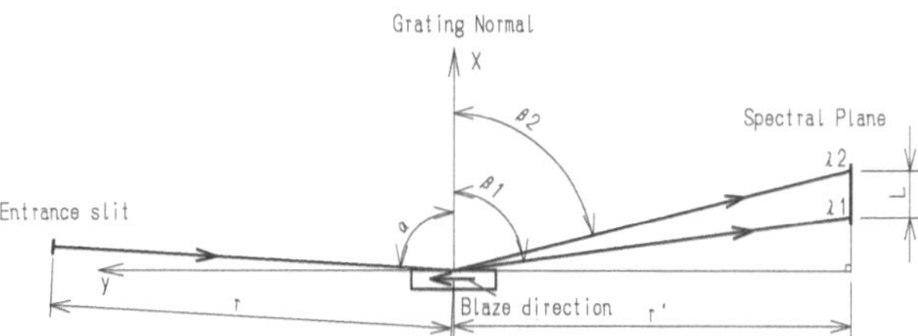

Figure 2.10: Schematic of VLS grating with the design parameters r (distance from slit to center of grating), r' (distance from center of grating to image plane), and L (the length of the spectral range that ...

Table 2.1: Parameters of Hitachi VLS gratings.

Part No.	Grooves per mm	Radius of curvature (mm)	Blaze WL between λ1 and λ2 (nm)	Blank size H×W×T (mm)	Blaze angle (degree)	α (degree)	r (mm)	β1 (degree)	β2 (degree)	r' (mm)	WL Range λ1 to λ2 (nm)
001-0437 *1,2	1200	5649	10	30×50×10	3.2	87	237	-83.04	-77.07	235.3	5~20
001-0266 *1,2	1200	5649	10	30×50×10	3.2	87	237	-83.04	-77.07	235.3	5~20
001-0450 *2	2400	15920	1.5	30×50×10	1.9	88.7	237	-85.81	-81.01	235.3	1~5
001-0471 *2	2400	15920	1.5	30×50×10	1.9	88.7	237	-85.81	-81.01	235.3	1~5
001-0639	600	5649	31	30×50×10	3.7	85.3	350	-79.56	-67.26	469	22~124
001-0640	1200	5649	16.3	30×50×10	3.78	85.3	350	-79.56	-67.26	469	11~62
001-0659 *3	2400	57680	3.3	40×70×12	3	89	564	-85.91	-80.21	563.2	1~6
001-0660 *3	1200	13450	9	40×70×12	3	87	564	-83.04	-75.61	563.2	5~25

References

1. T. Kita, T. Harada , N. Nakano and H. Kuroda, "Mechanically ruled aberration-corrected concave gratings for a flat-field grazing-incidence spectrograph", Appl. Opt. 22, 512-513 （1983）
2. N. Nakano, H. Kuroda, T. Kita and T. Harada, "Development of a flat-field grazing-incidence XUV spectrometer and its application in picosecond XUV spectroscopy", Appl. Opt. 23, 2386-2392 （1984）
3. T. Harada, K. Takahashi, H. Sakuma and A. Osyczka, "Optimum design of a grazing-incidence flat-field spectrograph with a spherical varied-line-space grating", Appl. Opt. 38, 2743-2748(1999)

and b_2, σ, σ_0 are given by

$$\sigma(w) = \frac{\sigma_0}{1 + \frac{2b_2}{R}w + \frac{3b_3}{R^2}w^2 + \frac{4b_4}{R^3}w^3 + ...} \tag{2.3}$$

which is the groove density along the face of the grating w [28–30].

Another design criteria for this spectrometer was that it had to be able to be used on a variety of different systems. The modular design of the TABLe meant that this spectrometer could be removed from the TABLe and brought to a different light source. This entails that the entrance arm could be different from the one specified by the manufacturer depending upon the system that is installed in. To account for this change in entrance arm, the angle of the grating can be changed from the specified one to recover the flat field condition. This can be seen for both gratings in figure 2.12. Of course, one issue with this approach is that even though the flat field can be recovered, the grating efficiency changes as a function of angle. This can be done numerically, and there are many packages available to perform these calculations. The grating efficiency for the 1200 lines/mm grating was calculated in [32], and the results are shown in figure 2.13. As can be clearly seen, the efficiency changes non-trivially as a function of grazing angle for a given wavelength. Thus, depending upon the position of the detector and the photon energy of interest, there is a different incident angle that can be selected to optimize the grating efficiency.

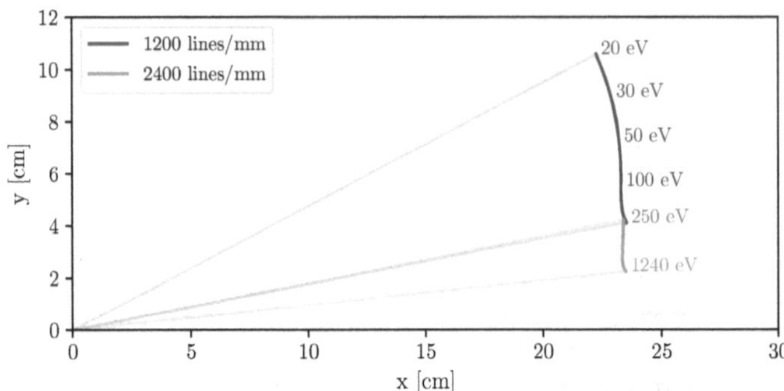

Figure 2.11: Flat fields of the two VLS gratings that comprise the photon spectrometer. The x axis is parallel to the face of the grating, and the y axis is perpendicular to the grating face. The center of the grating is located at (0,0). Incidence angle and entrance arm are those provided in table 2.1.

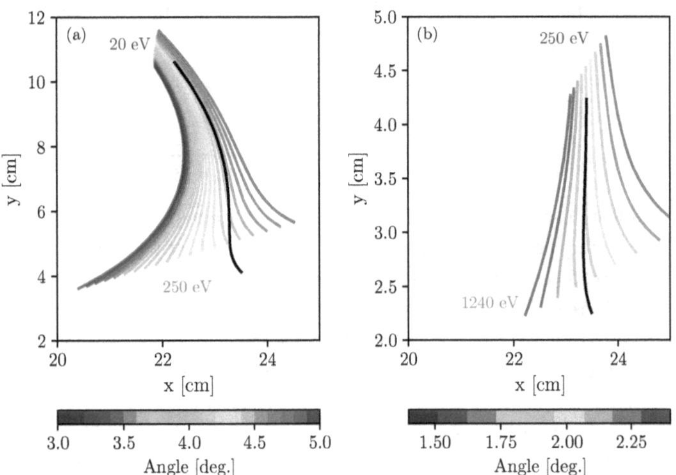

Figure 2.12: Flat field calculated as a function of incident angle for an entrance arm of 1.5 m. The flat field for the manufacturers specified entrance arm of 0.237 m is shown in black. (a) is for the 1200 lines/mm VLS grating and (b) is for the 2400 lines/mm VLS grating. The x axis is parallel to the incident light, and the y axis is perpendicular to the incident light.

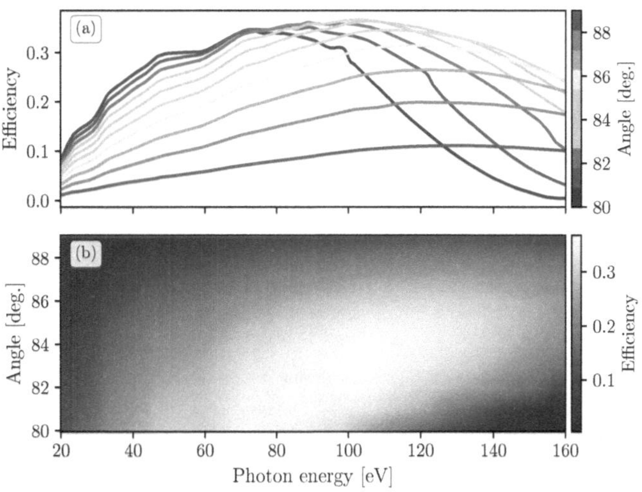

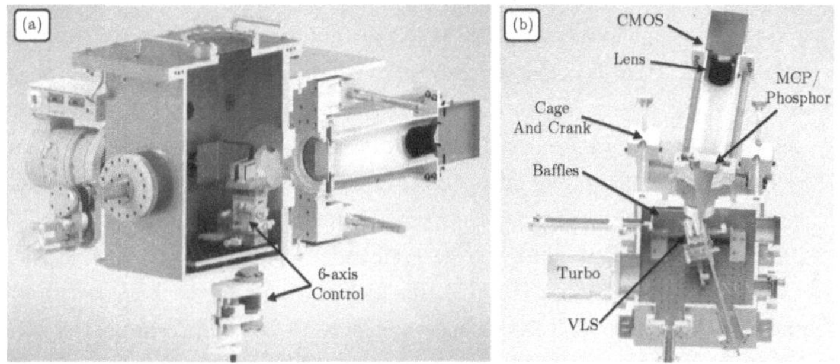

Figure 2.14: Render of the spectrometer showing a side-view (a) and a top-view (b).

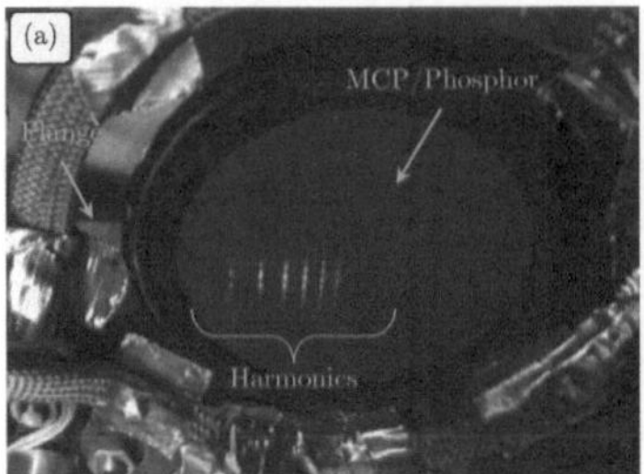

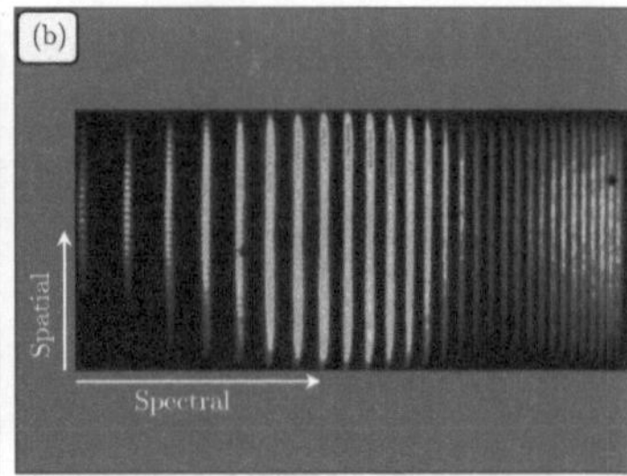

Figure 2.15: (a) Camera image of the output of the phosphor screen. Harmonics are visible by eye. (b) Sample data image collected from the Andor high-resolution camera. Spatial and spectral structure of the XUV APT can be seen.

Now it is time to move on to the actual design of the spectrometer, and a render of the spectrometer is shown in figure 2.14. This design has the capability to switch between the two diffraction gratings while under vacuum through the use of 6-axis motorized control. The detector consists of a 75 mm imaging quality micro-channel plate (MCP) and phosphor that is mounted on an 8" ConFlat flange. The output of the phosphor is re-imaged onto an Andor Neo 5.5 CMOS camera using a 50 mm focal length lens. The MCP was selected over a direct detection XUV CCD camera because of the much larger detector area that can be achieved using an MCP. To position this detector along the focal plane for each grating, a custom manipulator was developed, and this manipulator is referred to as the cage and crank. It consists of three screws that allow for the tilt and translation of the detector while the chamber is under vacuum. This allows for optimization of the resolution in different parts of the harmonic spectrum based upon the experimental needs. An example of the phosphor output is shown in figure 2.15 (a) for a case where the phosphor emission was bright enough that it could be easily seen by eye, and an example of a typical image taken by the Andor camera is shown in figure 2.15 (b). These images constitute the basis of all of the experiments that are described within this book, and there are a couple of key features that are important to highlight. As a consequence of selecting VLS gratings, the resulting spectrum that is measured is only focused along the spectral dimension, as shown in figure 2.15 (b), and the other dimension, generally referred to as the spatial dimension, is unfocused and allows for the measurement of the spatial profile of the XUV beam as a function of energy. This allows for unique measurements to be performed that involve the spatial interference of two XUV beams, and this capability is what enables the experiments in Chapters 3, 4, and 6.

Finally, the last important consideration of the spectrometer is its maximum achievable

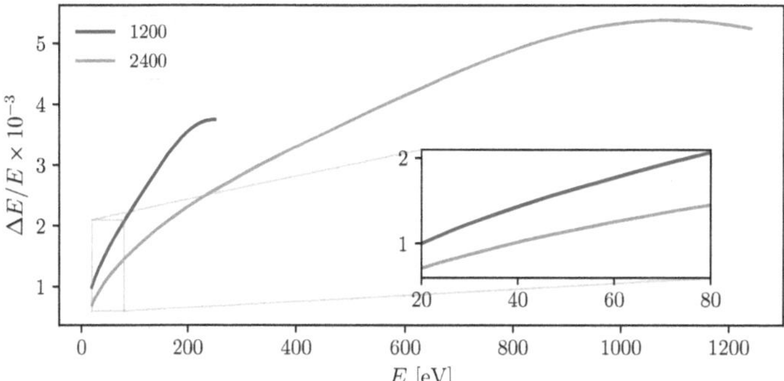

Figure 2.16: Maximum achievable resolution for the 1200 and 2400 lines/mm VLS gratings. Calculated using an effective pixel size of 48 μm. Inset shows the spectral region that will be important for the experiments described herein.

resolution. This can be done by calculating the energy difference between two points that are separated by the effective pixel size along the flat field curve. This calculation is shown in figure 2.16 for both gratings. In the energy range of 20-80 eV, the resolution $\Delta E/E$ of the 1200 lines/mm grating ranges from 1×10^{-3} to 2×10^{-3}. This resolution is fundamentally limited by the effective pixel size, and it can be improved by a factor of 4 if a different re-imaging lens is used for the Andor camera. Further resolution improvements can be made by using a direct detection XUV CCD camera with smaller pixel sizes.

2.4 Optical layout

Now that the design of the TABLe has been covered, attention can now be turned to the optical layout of the TABLe. As stated previously, the TABLe is designed around performing pump/probe experiments using both XUV and IR pulses. This is done by building a Mach-Zehnder interferometer where one arm is IR and the other arm is used to generate the XUV APT. The optical layout that was implemented is shown in figure 2.17. This setup was designed to use the signal wavelengths from the TOPAS over the 1250-1550 nm range, and it can accommodate a wavelength change within this range with minimal adjustment of the interferometer. In the following sections several major components that determine the interferometer performance will be highlighted, and the two different XUV generation schemes that are used will be described in Chapter 3 and 5.

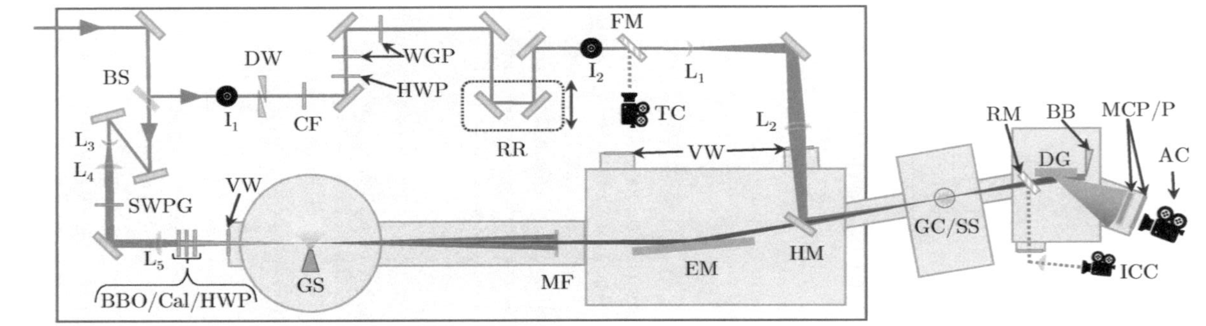

Figure 2.17: Schematic of the beam path for the TABLe interferometer. The input laser is shown in red, its second harmonic in blue, and the generated XUV in purple. **BS**: Beamsplitter (Thorlabs BSF20-C), I_1 and I_2: Irises used for alignment into interferometer. **DW**: Delay wedges for fine delay control, see section 2.4.1. **CF**: Color filter to remove parasitic colors from TOPAS (Thorlabs FELH1000). **HWP**: Half-wave plate. **WGP**: Wire grid polarizer. **RR**: Retro reflector for coarse delay adjustment. **FM**: Flip mirror. **TC**: Thermal camera used for alignment. L_1: $f = -300$ mm lens (Thorlabs LF1015-C). L_2: $f = 500$ mm lens (Thorlabs LA1380-C). **VW**: Vacuum window, 3 mm CaF_2, **HM**: Hole mirror with 10 mm hole. L_3: $f = -400$ mm lens. L_4: $f = 500$ mm lens. **SWPG**: Square-wave phase grating. L_5: $f = 400$ mm lens. **BBO**: Second-harmonic generation crystal. **Cal**: Calcite. **GS**: Gas source for HHG. **MF**: Metallic filter. **EM**: Ellipsoidal mirror. **GC/SS**: Gas cell or solid sample. **RM**: Removable mirror for *in-situ* diagnostics. **ICC**: camera for *in-situ* diagnostics. **DG**: VLS diffraction grating. **BB**: Baffles to block zero order diffraction. **MCP/P**: Microchannel plate and phosphor. **AC**: Andor Neo 5.5 CMOS camera.

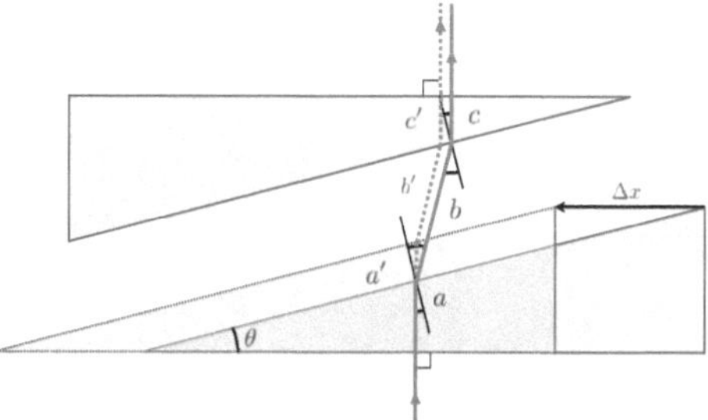

Figure 2.18: Schematic of the FS wedges used to control the time delay between the IR and XUV pulses in the dressing and generation arms interferometer, respectively. The wedges are aligned such that the input beam is normal to the first wedge face, and the beam exits the wedges normal to the last face of the second wedge. Only one of the wedges is motorized and is shown before and after a displacement by an amount Δx.

2.4.1 Time Delay Control

An important consideration in pump/probe experiments is how to control the delay between the pump and probe pulses. To control this delay, there are typically two methods that can be employed optically. The first to to use a retro-reflector that is mounted on a motorized stage [33–38]. With this method, the delay is simply related to the displacement of the motorized stage by the relationship $\Delta\tau = 2\Delta x/c$. This means that a displacement of 10 nm by the motorized stage would lead to a delay of 67 as, whereas a displacement of 2 in. would lead to a delay of 339 ps. This setup is advantageous if large delays are required (10s of ps to a few ns), however for the short time steps that are required for an attosecond measurement (typically on the order of 100 as) the mechanical requirements on the motor being used to move the retro are very high. Since a 100 as step would equate to a translation of 15nm, this would require the use of a piezoelectric motor. Piezo motors and their associated electronics tend to be expensive, and they have inherent problems because they exhibit nonlinear movement due to hysteresis and they tend to drift and creep after actuation. This can be abated through a feedback sensor and operating it in a closed-loop mode, however the quality of the sensor and electronics determines how effectively these problems are minimized.

The second common method to control the delay between the pump and probe pulse is to use a pair of glass wedges [21, 39, 40]. A schematic of how this is achieved is shown in figure 2.18. In this scheme, only one of the glass wedges is motorized, and the direction of translation is perpendicular to the input beam and parallel to the first glass face. The path that a ray would take through the wedge pair is shown before ($a \rightarrow b \rightarrow c$) and after

$(a' \rightarrow b' \rightarrow c')$ translation by an amount Δx. Assuming that the wedge angle is θ, the path after translation by Δx can be written as

$$a' = a + \Delta x \tan \theta \tag{2.4}$$

$$b' = b - \Delta x \left(\frac{\sin \theta}{\cos \psi} \right) \tag{2.5}$$

$$c' = c + \Delta x \tan \theta \left(\frac{\sin(\psi - \theta)}{\cos \theta} \right) \tag{2.6}$$

where

$$\psi = \arcsin(n \sin \theta) \tag{2.7}$$

is given by Snell's Law [27]. From the difference in optical path length between these two paths, one can calculate the time delay $\Delta \tau$ introduced by a translation of Δx, and this relationship is given by

$$\Delta \tau = \frac{\Delta x}{c} \left[n \tan \theta - \frac{\sin \theta}{\cos \psi} - n \tan \theta \left(\frac{\sin(\psi - \theta)}{\cos \theta} \right) \right]. \tag{2.8}$$

For a pair wedges made out of fused silica (Infrasil) with a wedge angle of $\theta = 4°$ and a beam of wavelength 1430 nm (n=1.4454), equation 2.8 becomes

$$\Delta \tau = \left(102 \left[\frac{\text{as}}{\mu\text{m}} \right] \right) \Delta x. \tag{2.9}$$

This entails that a translation of 1 μm would lead to a delay of only 100 as. This reduction in motor step to delay step ratio compared to the retroreflector case means that the requirements on the motorized stage are greatly reduced. Additionally, since the glass wedges are a transmissive optic, they are inherently less sensitive to vibrations when compared to a retroreflector.

In the TABLe apparatus, both types of delay control have been implemented, as shown in figure 2.17. The retroreflector is mounted on a translation stage with 2 inches of travel that is controlled manually with a micrometer. The primary use for the retroreflector is to make coarse adjustments to the dressing arm to account for changes in temporal overlap between the two arms of the interferometer. Typically, this is due to adjustments made to the interferometer itself (such as introducing new optics) or due to changes in the input laser, usually either pointing or wavelength.

To finely control the delay, a pair of glass wedges is used. These wedges are made out of fused silica, and have a wedge angle of $\theta = 4°$. The first of the two wedges are motorized in manner similar to that shown in figure 2.18. The stage that the first wedge is mounted to has a total travel of 1 inch, and it is controlled by a Thorlabs Z825B DC servo motor. This "pencil" motor, as it is known in the lab, has a minimum repeatable incremental motion of 0.2 μm and is encoded, so it's absolute position is known to within the homing accuracy of ± 1 μm. From equation 2.9, using these wedges at 1430 nm with a step size of 1 μm will give a delay of 101 as.

2.4.2 IR Dressing Intensity

An important consideration in any pump/probe experiment is the intensity of the IR field that is used as a probe/dressing field. Ideally, one would like to be able to control the intensity such that both perturbative and strong-field regimes can be accessed with the same optical setup. This can be achieved by selecting an optical setup that has a high peak intensity and then attenuate the beam to achieve a lower intensity. Attenuation can be achieved through the use of neutral density (ND) filters, however they can lead to several complications. Since the ND filter would be placed in only one arm of the interferometer to control only the dressing intensity, any variation in thickness between different ND filters would lead to a change in temporal overlap. Additionally, any change in positioning when switching ND filters will lead to a slight change in spatial overlap. In light of this, the method to attenuate the beam that was implemented in the TABLe interferometer is a half-wave plate (HWP) and a wire grid polarizer (WGP). By rotating the polarization using the HWP, we can finely control the intensity of the dressing field by using the WGP to transmit only one component of the rotated polarization. The WGP is set such that the initial polarization is maintained, and this insures that the polarization of the IR and XUV are parallel in the interaction region. A second WGP is used to increase the effective extinction ratio to ensure that the field is linearly polarized even when it is being strongly attenuated. The extinction ratio of one of the WGP (Thorlabs WP25M-UB) is approximately 12500:1 at 1430 nm. Using this setup in conjunction with a beam splitter that reflects 8% of the input energy to the dressing arm, the pulse energy can be tuned between 1 μJ and 125 μJ.

The next design consideration for the dressing arm of the interferometer is choosing focusing optics to set the peak intensity that is achievable at the interaction region. The two main options in this regard are either focusing mirrors or lenses. Focusing mirrors have a significant advantage in the fact that they are achromatic, however their use leads to an optical system that is generally larger in optical path length and is difficult to switch between focal geometries. Lenses in that regard are well suited to adjusting between different experimental configurations without changing the overall footprint of the interferometer. The dressing lenses that are used in all of the experiments in this book are shown in 2.19, and they were originally selected to achieve the highest possible intensity at the focal plane given the geometrical constraints of the TABLe [21]. This choice of lenses was made to drive strong-field processes in rare gas atoms in the interaction region, and they were selected based upon calculations done by D. Kiesewetter using a hole mirrors as both beam splitters in the interferometer [21]. His calculations estimated that the peak intensity

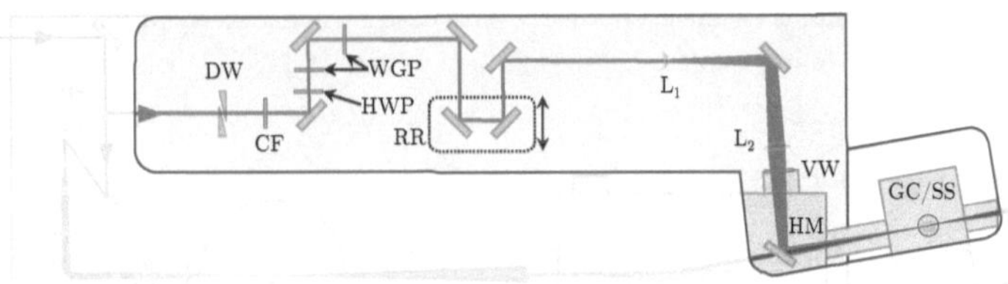

Figure 2.19: Optical layout of the dressing arm of the TABLe interferometer. HWP and WGP are used to finely control the power in the dressing arm. The lenses L_1 and L_2 were chosen to achieve a high intensity at the focal plane in the target chamber. This was done to have the capability to drive strong-field processes in a gas medium [21]. See figure 2.17 for full details of the interferometer.

at 1300 nm was 0.59 TW/cm^2 for a pulse energy of 1 μJ just before L_1 in figure 2.19.

For the experiments described herein, the initial beam splitter was changed to a beam sampler, and the recombining hole mirror was changed to a hole diameter of 10 mm from 6 mm. This was done to maximize the XUV flux by sending as much energy to generation as possible while minimizing clipping of the XUV with the hole mirror. In light of these changes, it is important to determine what the new peak intensity is at the focus. To accurately determine the intensity, a beam propagation simulation is performed using the measured beam profile. It is important to use the measured beam profile because the spatial mode out of the TOPAS is decidedly non-Gaussian, and this can be seen in figure 2.20 (a) which shows the beam profile at 1430 nm measured on a thermal camera. Beam propagation is implemented using an open-source Python package developed by Flexible Optical B.V. (OKO Tech) called LightPipes. The package consists of numerical methods to calculate the Fresnel integral

$$u(x, y, z) = \frac{ik}{2\pi z} e^{ikz} e^{ik(x^2+y^2)/2z} \int_{-\infty}^{\infty} \int_{-\infty}^{\infty} u(\xi, \eta, 0) e^{ik(\xi^2+\eta^2)/2z} e^{-ik(x\xi+y\eta)/z} \, d\xi \, d\eta \quad (2.10)$$

which gives the field $u(x, y, z)$ after propagation of a distance z, given an initial field profile of $u(x, y, 0)$. Using this formalism, a thin lens of focal length can be treated as adding a quadratic phase $\phi = -\pi(x^2 + y^2)/\lambda f$ to the field, and the effect of apertures can also be included to account for the effect of diffraction on the beam profile and intensity [41].

The results of the simulation are shown in figure 2.20. This simulation was performed for an input pulse energy of 1 μJ at a wavelength of 1430 nm. The simulation begins just before L_1 in 2.19 and the beam is propagated through to the interaction region GC/SS. The

26

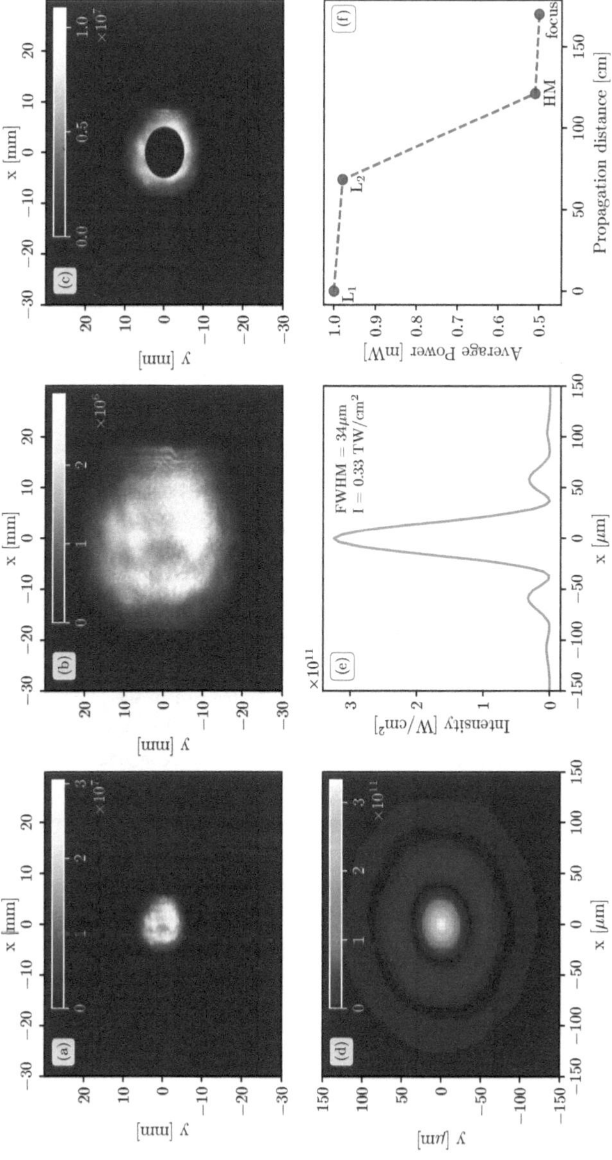

Figure 2.20: (a) Beam profile of the signal from the TOPAS pumped by the Spitfire laser system. Wavelength is 1430 nm and the pulse is normalized to 1 µJ of pulse energy. Beam profile was measured using a thermal camera, and is used as the input to the beam propagation simulation that is initiated just before L_1 in figure 2.19. (b) Numerically propagated beam profile just after L_2 in figure 2.19. (c) Beam profile just after reflecting off the hole mirror. (d) Calculated intensity profile at the focal plane in the interaction region. (d) Lineout of the calculated intensity profile. For an input pulse energy of 1 µJ, a peak intensity of 0.33 TW/cm^2 can be achieved. (f) Integrated average power at different points in the calculation. The main source of energy loss is due to the hole mirror, which only reflects 52% of the incident light. The hole mirror has an inner radius of 5 mm in this case.

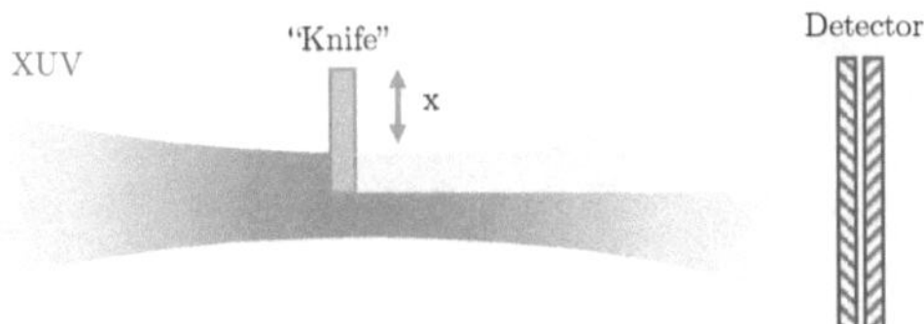

Figure 2.21: Schematic of knife-edge technique to measure beam size. By translating a sharp beam block through the focus and measuring the total transmitted power the beam profile can be reconstructed.

beam profile at L_2 and immediately after the hole mirror is shown in figure 2.20 (b) and (c), and the intensity profile at the focus is shown in figure 2.20 (d). From these simulations, the peak intensity for a pulse energy of 1 μJ is 0.33 TW/cm^2. For the range of pulse energies available in the dressing arm, this corresponds to an intensity range of 0.33 - 41.3 TW/cm^2 that is achievable.

2.4.3 XUV Beam Size

Another important consideration in any pump/probe experiment is the beam size of the probe relative to the pump in the interaction region. The reason for this is because the probe beam inherently samples the excited state of the system induced by the pump at intensities within the intersection of the pump and probe focal volumes contained within the system being studied. Thus, a smaller probe waist relative to the pump waist means that a narrower distribution of pump intensities will be sampled by the XUV probe. The width of this distribution limits the ability to study intensity dependent effects, and it is a critical parameter to control. As stated previously, the demagnification of the XUV spot size provided by the ellipsoidal mirror helps tremendously to achieve as small ratio of probe to pump beam sizes even at small pump beam sizes that are used.

Additionally, it is also of great importance to place the sample at the focus of both the XUV and the IR beams in the target chamber. This can be done easily with the IR either through direct imaging of the beam or by using an intensity dependent effect (such as second harmonic generation from a BBO) to find the peak intensity along the k-direction of the IR beam. The focus of the XUV is more difficult to find because it must be found in vacuum. One method to find the focus is to measure the beam diameter at various points along the k-direction of the XUV to find the minimum beam diameter. A method to do this is to employ a knife-edge measurement [42–45].

The principle behind the knife-edge measurement is simple, and it is shown schematically in figure 2.21. In the measurement, a "knife" is translated through the beam to be measured perpendicular to the k-direction of the beam. The knife is simply an opaque material that has a sharp edge to minimize scattered light. As this knife is translated through the focus, the total power is measured further downstream as a function of the knife's position within the beam, and this dependence is given by

$$P(x,z) = \int_{-\infty}^{\infty} \int_{-\infty}^{x} I(x', y', z)\,\mathrm{d}y'\,\mathrm{d}x' \tag{2.11}$$

where $I(x, y, z)$ is the intensity of the beam. For a Gaussian beam, this relationship becomes

$$P(x,z) = \frac{P_0}{2}\mathrm{erfc}\left(\frac{x\sqrt{2}}{w(z)}\right) \tag{2.12}$$

where $w(z)$ is waist radius as a function of z and is given by

$$w(z) = w_0\sqrt{1 + \left(\frac{z - z_0}{z_R}\right)^2} \tag{2.13}$$

where $z_R = \pi w_0^2 n/\lambda$ is the Rayleigh range and z_0 is the position of the focus along the k-direction.

An example of this technique being used to measure the beam size of the XUV is shown in figure 2.22. In this case, the knife being used is the beveled edge of a silicon frame that is 300 μm thick. The thickness of the frame is such that the transmission is negligible through the frame itself. The integrated harmonic signal is fit to

$$P(x) = \frac{a}{2}\mathrm{erfc}\left(\frac{\sqrt{2}(x - x_0)}{w}\right) + b, \tag{2.14}$$

and the beam size can be extracted from the fit. In this case the beam size was 9 μm at this position of the XUV focus. Comparing this beam size to the predicted beam size of the IR, we can see that the XUV spot size relative to the IR spot size is roughly three times smaller.

This measurement can repeated along the k-direction of the XUV beam, and the focus can be mapped out. An example of this type of measurement is shown in figure 2.23. As can be seen, the fit to a Gaussian agrees quite well with the measure dependence of the beam profile. The extracted Rayleigh range is 1.1 mm with a waist of 6.1 μm at a motor position of 12.7 mm along the k-direction. This type of measurement is performed whenever the optical setup is changed in the generation arm of the interferometer, as it insures that the sample is always at the focus of the XUV.

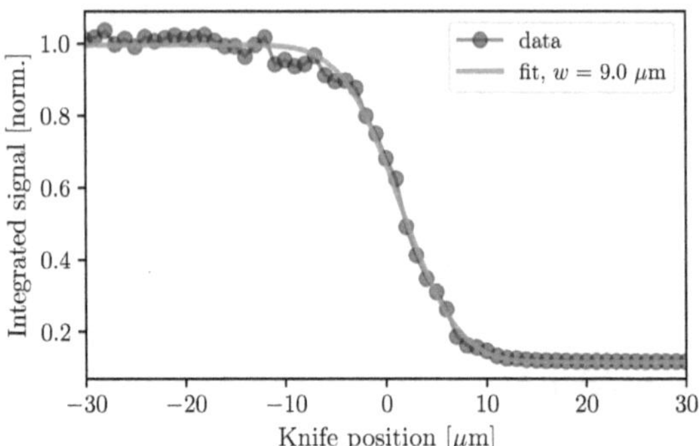

Figure 2.22: Schematic of knife-edge technique to measure beam size. By translating a sharp beam block through the focus and measuring the total transmitted power the beam profile can be reconstructed.

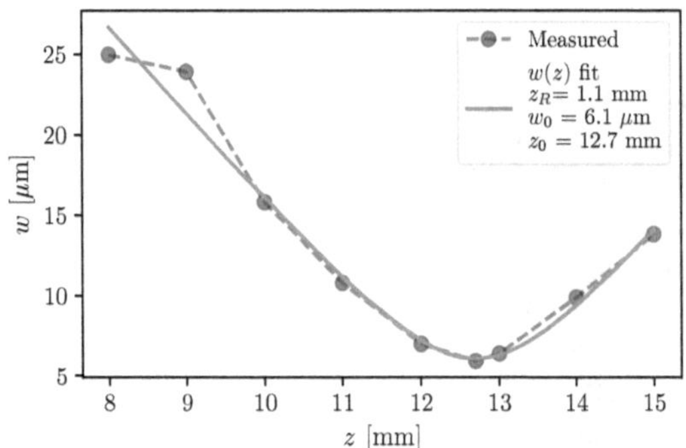

Figure 2.23: Focus of the XUV measured using knife-edge method at various points along the k-direction of the XUV beam.

2.5 Conclusion

In this chapter the design and optical layout of the Transient Absorption Beamline (TABLe) was described. This experimental apparatus represent the culmination of years of design, construction, and validation performed by Greg Smith and myself. The unique capabilities of the TABLe enabled the work detailed in this book to be performed, as well as other experiments [21, 46, 47].

Chapter 3
Two-source high harmonic generation

3.1 Introduction

A common difficulty in working with extreme ultraviolet (XUV) light is the lack of efficient and broadband optics, especially beam splitters. In this chapter, I will introduce a method for generating two sources of XUV light by high harmonic generation (HHG) using a square-wave phase grating (SWPG). This SWPG allows for the duplication of an infrared (IR) pulse, as well as precise and stable control of the relative phase between the duplicates of the input IR pulse. The two most intense duplicates can generate harmonics which will interfere in the far-field. This can be thought of as an inline Mach-Zehnder interferometer with interferometric stability on sub-wavelength level of the high harmonic. The inherent stability of this two-source scheme will be utilized to measure both the real and imaginary parts of the refractive index of a medium.

3.2 Theory

3.2.1 Laser beam shaping using diffractive optics

In many experimental designs, it is advantageous to be able to shape the spatial intensity distribution of light to be something other than a typical Gaussian beam. A common example of this is generating a beam with an approximately constant intensity across its spatial profile (a flat-top beam). For the experiments described herein, we will be interested in duplicating an input beam with relative phase control between the two replica beams. Both of these examples are part of the general concept of laser beam shaping. The challenge is to design an optical system such that given an input beam profile $I_{in}(x, y)$ we can generate the desired output beam profile $I_{out}(x, y)$. Ideally, this optical system is designed in such a way that it can be nearly lossless. The relevant optical system which will be discussed in this chapter is shown in figure 3.1. The system consists of a phase element which modifies

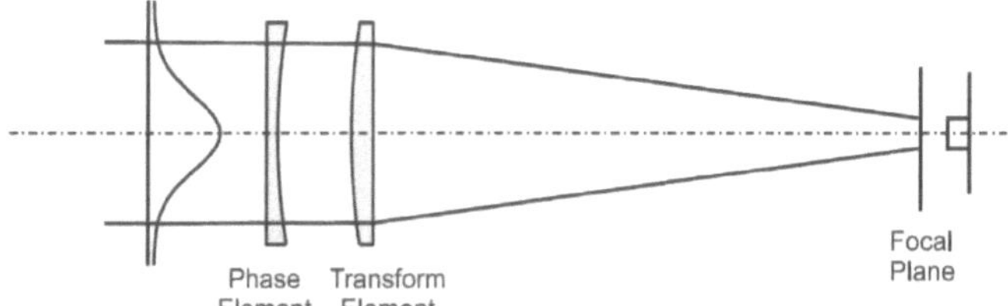

Figure 3.1: Schematic demonstrating how to use a diffractive optical element to shape the beam profile at the focal plane. A collimated coherent beam is incident upon a diffractive optical element which shapes the phase of the incident beam, and then a lens is used as a transform element to Fourier transform the beam at the focal plane. The intensity profile at the focal plane can be controlled by altering the spatial dependence of the phase imparted upon the incident beam by the phase element. Adapted from [18]

the phase of the input field by $\phi(x, y)$ and a Fourier transform lens which adds a quadratic phase to the beam to focus it at the focal plane. By appropriate choice of the phase profile of the phase element, one can produce the desired beam profile at the focal plane.

This problem can be theoretically described in terms of Fourier optics [11, 48, 49]. If one assumes that a field $u(x, y, 0)$ is incident upon an aperture at $z = 0$, then the field for $z > 0$ can be written under the Fresnel approximation by the integral

$$u(x, y, z) = \frac{ik}{2\pi z} e^{ikz} e^{ik(x^2+y^2)/2z} \int_{-\infty}^{\infty} \int_{-\infty}^{\infty} u(\xi, \eta, 0) e^{ik(\xi^2+\eta^2)/2z} e^{-ik(x\xi+y\eta)/z} \, d\xi \, d\eta \quad (3.1)$$

where $u(\xi, \eta, 0)$ is the incoming field and $k = 2\pi/\lambda$ is the wavenumber. Now, if one assumes that the phase element is placed at $z = 0$, then immediately after passing through the thin phase element in Fig. 3.1 the field is given by

$$u(\xi, \eta, 0) = F(\xi, \eta) e^{i\phi(\xi, \eta)}. \quad (3.2)$$

After propagating through the thin Fourier transform lens of focal length f, a phase of $k(\xi^2 + \eta^2)/2f$ is added to the beam, and the field at the focal plane is now given by

$$u(x, y, f) = \frac{ik}{2\pi f} e^{ikf} e^{ik(x^2+y^2)/2f} \int_{-\infty}^{\infty} \int_{-\infty}^{\infty} F(\xi, \eta) e^{i\phi(\xi, \eta)} e^{-ik(x\xi+y\eta)/f} \, d\xi \, d\eta \quad (3.3)$$

where the quadratic phase in the integral of equation 3.1 is exactly canceled by the quadratic phase introduced by the lens.

Now, the idea is to rewrite this field profile at the focal plane into a more intuitive form by introducing the equation

$$g(\xi, \eta) = \frac{ik}{2\pi f} F(\xi, \eta) e^{i\phi(\xi, \eta)}.$$

(3.4)

The Fourier transform of this function $g(\xi, \eta)$ is given by

$$G(a, b) = \frac{ik}{2\pi f} \int_{-\infty}^{\infty} \int_{-\infty}^{\infty} F(\xi, \eta) e^{i\phi(\xi, \eta)} e^{-i(a\xi + b\eta)} \, d\xi \, d\eta.$$

(3.5)

By setting $a = kx/f$ and $b = ky/f$ and taking the square complex modulus, one obtains the equation

$$|G(kx/f, ky/f)|^2 = \frac{k^2}{(2\pi f)^2} \left| \int_{-\infty}^{\infty} \int_{-\infty}^{\infty} F(\xi, \eta) e^{i\phi(\xi, \eta)} e^{-ik(x\xi + y\eta)/f} \, d\xi \, d\eta \right|^2.$$

(3.6)

By comparing this equation with the square complex modulus of the field at the focal plane (equation 3.3), one finds the relationship

$$|u(x, y, f)|^2 = |G(kx/f, ky/f)|^2.$$

(3.7)

From this last equality we have shown that the intensity of the field at at the focal plane $|u(z = f)|^2$, is given by the Fourier transform of the combined phase imparted upon the incident field by both the phase element and the Fourier transform lens, $|G|^2$. So, if one wants a specific beam shape $Q(x, y)$ at the focal plane, then by tuning the frequency components of the phase imparted upon the beam $\phi(x, y)$ and the focal length f used then one can achieve the desired beam profile, such that

$$|G(kx/f, ky/f)|^2 = Q(x, y).$$

(3.8)

This problem is difficult in general, so the challenge in many beam shaping problems is to try and minimize the error between the actual beam profile and the desired profile, and, to further complicate the matter, many applications will require different notions of error to be used. For example, if one needs the energy distribution to be as close as possible to the desired profile, then the ℓ_2-norm would be appropriate. However, if the maximum intensity is of concern, then the ℓ_∞-norm combined with the ℓ_2-norm would be the appropriate notion of error.

It is possible to gain more insight into how difficult a beam shaping problem will be by reformulating the problem in terms of relevant length scales [48, 49]. The idea is to introduce a dimensionless parameter whose magnitude will reflect the validity of underlying assumptions, and so for a given value of this parameter one can intuitively understand the performance (or lack thereof) of the beam shaping system. This is done by reformulating the above situation in terms of the natural length scales of both the incoming field and the desired field at the focal plane

$$I_{\text{input}} = |f(x,y)|^2 = |\hat{f}(x/\sigma, y/\sigma)|^2 \tag{3.9}$$

$$I_{\text{desired}} = Q(x,y) = \hat{Q}(x/D, y/D) \tag{3.10}$$

where σ is the characteristic length scale of the input field (typically the beam radius) and D is the characteristic length scale of the desired field. By expressing the fields in this way, we can now introduce the dimensionless parameter

$$\beta = \frac{2\pi\sigma D}{\lambda f}. \tag{3.11}$$

Using this dimensionless parameter, one can rewrite equations 3.5 and 3.8 as

$$G(\chi,\nu) = \int_{-\infty}^{\infty}\int_{-\infty}^{\infty} g(\xi,\eta) e^{-i(\chi\xi+\nu\eta)} e^{i\beta\hat{\phi}(\xi,\eta)} \, d\xi \, d\eta \tag{3.12}$$

$$|G(\chi,\nu)|^2 = \frac{4\pi^2 A}{\beta^2} Q(\chi/\beta, \nu/\beta) \tag{3.13}$$

where $\chi = x\sigma k/f$, $\nu = y\sigma k/f$, A is a constant, and $\hat{\phi} = \beta\phi$. From equation 3.12, it is clear that the functions g and G are related by a Fourier transform, so they must obey the uncertainty relation given by

$$\mu_g \mu_G \geq 1 \tag{3.14}$$

where μ is the second moment. Now, if one were to choose the phase profile of the phase element $\phi(x,y)$ such that equation 3.13 is satisfied, then one finds that $\mu_G = \beta^2 \mu_Q$. This then leads to the inequality

$$\beta^2 \mu_g \mu_Q \geq 1. \tag{3.15}$$

It can be seen that for large vales of β this inequality can be readily satisfied. However, for very small values of β this inequality cannot be met and it will not be possible to produce the desired beam profile. From this, it can be seen that having a large value of β makes the beam shaping problem more tractable. The physical interpretation of β is that is a measure of validity of geometric optics. It can be shown that when β is large a stationary phase method can be used to expand equation 3.12, and the lowest order term can be derived using a geometric optics approximation [18, 49].

In this section, the general problem of laser beam shaping has been introduced and formulated as a problem in Fourier analysis, and by rewriting everything in terms of natural length scales we can infer which types of beam shaping problems will be more tractable using geometrical optics approximations. The discussion so far has been kept very abstract, but in the next section the problem of interest will be introduced and these ideas will become more concrete.

3.2.2 Beam splitting phase grating

Now that the general theory behind laser beam shaping has been introduced, we move on to the specific problem at hand. The idea is to produce two nearly identical XUV beams through high-harmonic generation (HHG) using two nearly identical IR beams. The challenge is how to produce two nearly identical IR beams while minimizing the energy lost in the process. An additional requirement is that we can control the relative phase between these two IR beams. All of these requirements can be met through the use of a particular beam splitting phase grating [46, 50–52].

As shown in section 3.2.1, the beam shape at the focal plane of a lens can be controlled through appropriate choice of a phase element and the spatially dependent phase $\phi(x, y)$ which is imparted upon the incoming beam. We will still consider the schematic shown in figure 3.1. However, the phase element which will be considered in this section (the beam splitting phase grating) will only modify the phase in one dimension, $\phi(x, y) = \phi(x)$, and it will be a periodic function with a period of d, $\phi(x) = \phi(x + d)$. If we expand the function $P(x) = e^{i\phi(x)}$ in a Fourier expansion

$$P(x) = \sum_{n=-\infty}^{\infty} a_n e^{\frac{i2\pi n x}{d}} \tag{3.16}$$

$$a_n = \frac{1}{d} \int_{-d/2}^{d/2} P(\tilde{x}) e^{\frac{-i2\pi n \tilde{x}}{d}} \, d\tilde{x}, \tag{3.17}$$

then it can be shown that each of the Fourier coefficients represents a diffracted beam and the energy contained in each diffracted beam is given by the square complex modulus of the Fourier coefficient $|a_n|^2$ [50]. Thus, by making our phase element in figure 3.1 a phase grating, we have split the beam into many diffraction orders. We have defined the period of the phase grating by requiring $\phi(x) = \phi(x + d)$, however we have not yet determined its shape. This can be accomplished by specifying the distribution of energy into the various diffraction orders. Since we want to split the incoming beam into two duplicate beams, we are searching for d-periodic function that puts equal energy into two diffraction orders and a maximal amount of the input energy is put into those two orders. It can be shown that the d-periodic function which meets these criteria is a $0 - \pi$ square-wave phase grating given by

$$P(x, x_0) = \text{sgn}\left(\cos\left(\frac{2\pi(x - x_0)}{d} \right) \right) \tag{3.18}$$

where x_0 is an offset of position of the SWPG in the plane transverse to the optical axis [46, 52]. From this equation, it can be seen that the phase of the incoming beam is modulated by either 0 or π by the phase grating, and this is shown in figure 3.2 for the offset positions $x_0 = 0$ and $x_0 = d/10$.

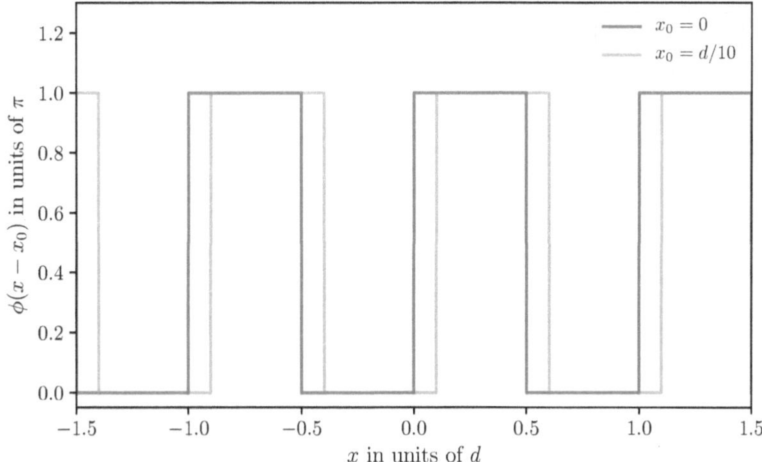

Figure 3.2: Plot of the phase function $\phi(x-x_0)$ in units of π for a $0-\pi$ SWPG with a period of d. The dark blue (light blue) curve shows the phase function for $x_0 = 0$ ($x_0 = d/10$).

From equation 3.16, we can calculate the Fourier coefficients $a_n(x_0)$ for the SWPG for $n = 0$ and $n \neq 0$. These Fourier coefficients determine how the energy is distributed between the different diffraction orders. For the zeroth-order case ($n = 0$), we find that

$$a_0(x_0) = \frac{1}{d} \int_{-d/2}^{d/2} \mathrm{sgn}\left(\cos \left(\frac{2\pi(\tilde{x} - x_0)}{d} \right) \right) \mathrm{d}\tilde{x}$$

$$a_0(x_0) = \frac{1}{d} \left[-\int_{-\frac{d}{2}}^{-\frac{d}{4}+x_0} \mathrm{d}\tilde{x} + \int_{-\frac{d}{4}+x_0}^{\frac{d}{4}+x_0} \mathrm{d}\tilde{x} - \int_{\frac{d}{4}+x_0}^{\frac{d}{2}} \mathrm{d}\tilde{x} \right] \tag{3.19}$$

$$a_0(x_0) = 0.$$

This demonstrates that zero energy is put into the zeroth-order diffraction for the $0 - \pi$

SWPG. For the other diffraction orders, we find that

$$a_n(x_0) = \frac{1}{d}\int_{-d/2}^{d/2} \mathrm{sgn}\left(\cos\left(\frac{2\pi(\tilde{x}-x_0)}{d}\right)\right)e^{-i\frac{2\pi n\tilde{x}}{d}}\,\mathrm{d}\tilde{x}$$

$$= \frac{1}{d}\left[-\int_{-\frac{d}{2}}^{-\frac{d}{4}+x_0} e^{-i\frac{2\pi nx}{d}}\,\mathrm{d}\tilde{x} + \int_{-\frac{d}{4}+x_0}^{\frac{d}{4}+x_0} e^{-i\frac{2\pi nx}{d}}\,\mathrm{d}\tilde{x} - \int_{\frac{d}{4}+x_0}^{\frac{d}{2}} e^{-i\frac{2\pi nx}{d}}\,\mathrm{d}\tilde{x}\right]$$

$$= \frac{1}{i2\pi n}\left[e^{\frac{i\pi n}{2}-\frac{i2\pi n x_0}{d}} - e^{in\pi} + e^{\frac{i\pi n}{2}-\frac{i2\pi n x_0}{d}} - e^{-\frac{i\pi n}{2}-\frac{i2\pi n x_0}{d}} + e^{-in\pi} - e^{-\frac{i\pi n}{2}-\frac{i2\pi n x_0}{d}}\right]$$

$$= \frac{1}{i\pi n}\left[e^{\frac{i\pi n}{2}-\frac{i2\pi n x_0}{d}} - e^{-\frac{i\pi n}{2}-\frac{i2\pi n x_0}{d}}\right] = \frac{\sin(n\pi/2)}{n\pi/2}e^{-i\frac{2\pi n x_0}{d}}$$

$$a_n(x_0) = \mathrm{sinc}(\frac{n\pi}{2})e^{-in\frac{2\pi x_0}{d}}.$$

$$(3.20)$$

Since $\mathrm{sinc}(n\pi/2) = 0$ for even integers n, we see that only the odd orders of diffraction from the SWPG are populated. The distribution of energy between the different diffraction orders is plotted in figure 3.3, and from this figure it is immediately clear that our choice of the $0 - \pi$ SWPG has succeeded in putting most of the input energy equally into two diffraction orders, namely the ± 1 orders. The efficiency of this phase grating can be defined as

$$\eta = |a_1|^2 + |a_{-1}|^2 = \frac{8}{\pi^2} \approx 0.8106, \tag{3.21}$$

which means that approximately 81% of the input energy will be put into the two orders that we want. It should be noted that $\sum_n |a_n|^2 = 1$, which means that while we can't get perfect conversion of energy into only two orders, this is still a lossless design.

With these Fourier coefficients in hand, we can now calculate the field profile at the focus. This is done by using equation 3.3,

$$\tilde{S}(\tilde{x}) = \frac{ikA}{2\pi f}e^{ikf}e^{i\frac{k\tilde{x}^2}{2f}}\int_{-\infty}^{\infty} S(x,x_0)e^{-ikx\tilde{x}/f}\,\mathrm{d}x$$

$$\tilde{S}(\tilde{x}) = \frac{ikA}{2\pi f}e^{ikf}e^{i\frac{k\tilde{x}^2}{2f}}\int_{-\infty}^{\infty} E(x)\sum_{n=-\infty}^{\infty} a_n(x_0)e^{-i\frac{2\pi nx}{d}}e^{-ikx\tilde{x}/f}\,\mathrm{d}x$$

$$\tilde{S}(\tilde{x}) = \frac{ikA}{2\pi f}e^{ikf}e^{i\frac{k\tilde{x}^2}{2f}}\sum_{n=-\infty}^{\infty} a_n(x_0)\int_{-\infty}^{\infty} E(x)e^{-i\frac{2\pi x}{\lambda f}(\tilde{x}-n\frac{\lambda f}{d})}\,\mathrm{d}x$$

$$\tilde{S}(\tilde{x}) = \sum_{n=-\infty}^{\infty} a_n(x_0)\tilde{E}(\tilde{x}-\tilde{x}_n)$$

$$(3.22)$$

where

$$\tilde{E}(\tilde{x}) = \frac{ikA}{2\pi f}e^{ikf}e^{i\frac{k\tilde{x}^2}{2f}}\int_{-\infty}^{\infty} E(x)e^{-ikx\tilde{x}/f}\,\mathrm{d}x, \tag{3.23}$$

$x_n = n\lambda f/d$, $S(x,x_0) = E(x)P(x,x_0)$ is the field after the phase grating, and A is a constant

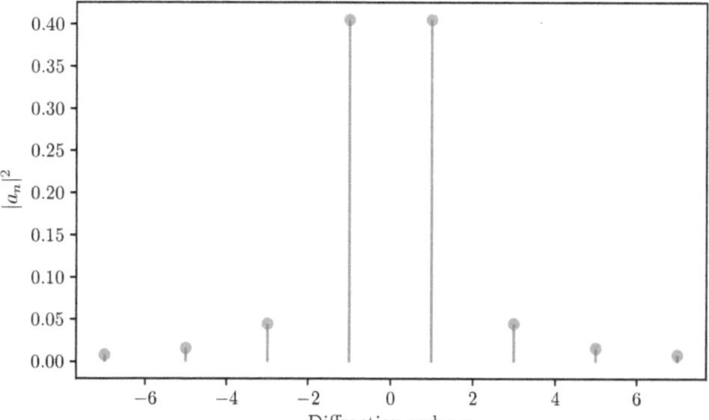

Figure 3.3: Square complex modulus of the Fourier coefficients a_n of the $0 - \pi$ SWPG. The square complex modulus is proportional to the energy put into each diffraction order. As can be seen from figure, the ± 1 orders have the most energy put into them at $4/\pi^2 \approx 41.1\%$ each. All even orders have zero energy.

to account for the y dimension in equation 3.3 which has been neglected for clarity in this discussion. Substituting in equation 3.20 into equation 3.22 yields

$$\tilde{S}(\tilde{x}, x_0) = \sum_{n \neq 0} \text{sinc}(\frac{n\pi}{2})\tilde{E}(\tilde{x} - \tilde{x}_n)e^{-in\frac{2\pi x_0}{d}} \tag{3.24}$$

which is the field at the focal plane. From this equation, the role of the transverse offset x_0 immediately becomes clear. It is used to control the relative phase between diffraction orders of the SWPG. The phase difference between the two most populated orders, the $n = \pm 1$ orders, is given by

$$\Delta\phi_{\pm 1} = 2\left(\frac{2\pi x_0}{d}\right). \tag{3.25}$$

Therefore, by controlling the offset of the SWPG we can control the relative phase between the two orders of interest over a range of $[0, 4\pi]$, two full periods of the fundamental wavelength. Additionally, one can begin to see from equation 3.3 that the intensity profile at the focal plane consists of copies of the focused input field at $x_n = n\lambda f/d$. An example intensity profile at the focal plane is shown in figure 3.4. In this figure the ± 1 orders are shown to be the most intense, and the phase is extracted for two different grating offset positions x_0. This is calculated by numerically propagating the beam profile and SWPG shown in figure 3.5.

So far, we have demonstrated that a binary $0 - \pi$ SWPG can theoretically achieve

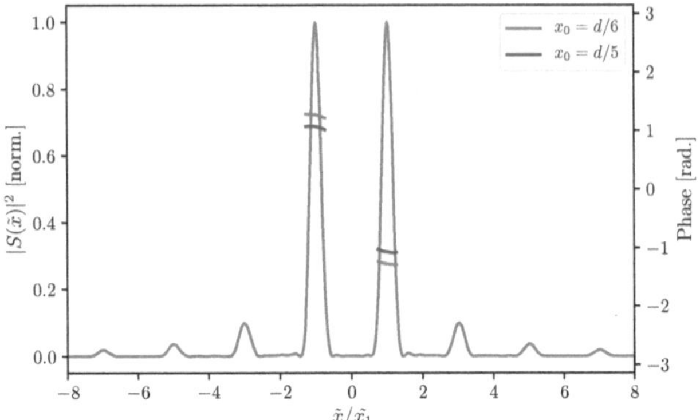

Figure 3.4: Intensity profile $S(\tilde{x})$ at the focal plane. Horizontal units are scaled by the spacing between orders, $\tilde{x}_1 = \lambda f/d$. Phase is also plotted for two different offset positions $x_0 = d/3$ and $x_0 = d/5$. This demonstrates the ability of the SWPG to generate two sources and control the relative phase between them. Calculated by numerically propagating the beam profile and SWPG in figure 3.5.

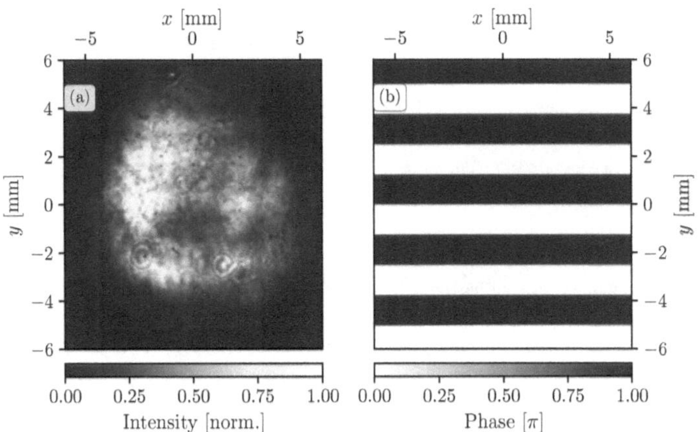

Figure 3.5: (a) Intensity profile of the input beam measured by a thermal camera. (b) Phase imparted by $0 - \pi$ SWPG with a grating period of $d = 2.5$ mm.

our requirements of an efficient beam duplicator with phase control between the two duplicate beams. However, all of the results shown above have been only considering the monochromatic case, and for the experiments of interest we use a femtosecond IR pulse with bandwidth on the order of 50 nm. This presents a challenge because the SWPG will be constructed by etching a fused-silica plate to have the desired phase step of π, and the inherent dispersion as the beam passes through the material means that the step will be π for only one wavelength. Thus, it is important to get a handle on how an imperfect non-π phase step affects the properties of the SWPG. To do this, we introduce a non-π phase step into the above analysis by a parameter ζ, such that the $0 - \pi$ step becomes a $0 - \zeta\pi$ step,

$$\phi(x, x_0, \zeta) = \zeta\phi(x, x_0). \tag{3.26}$$

Going back to 3.19, we can calculate the zero-order term for $\zeta \neq 1$,

$$\begin{aligned}
a_0(x_0, \zeta) &= \frac{1}{d} \int_{-\frac{d}{2}}^{\frac{d}{2}} e^{i\zeta\phi(\tilde{x},x_0)} \, \mathrm{d}\tilde{x} \\
&= \frac{1}{d} \left[\int_{-\frac{d}{2}}^{-\frac{d}{4}+x_0} e^{i\zeta\pi} \, \mathrm{d}\tilde{x} + \int_{-\frac{d}{4}+x_0}^{\frac{d}{4}+x_0} \mathrm{d}\tilde{x} + \int_{\frac{d}{4}+x_0}^{\frac{d}{2}} e^{i\zeta\pi} \, \mathrm{d}\tilde{x} \right] \\
&= \frac{1}{d} \left[\frac{d}{2} + \frac{d}{2}e^{i\zeta\pi} \right] = \frac{e^{i\zeta\pi/2}}{2} \left[e^{i\zeta\pi/2} + e^{-i\zeta\pi/2} \right] \\
a_0 &= \cos\left(\frac{\pi}{2}\zeta\right) e^{i\zeta\pi/2}.
\end{aligned} \tag{3.27}$$

Previously, for $\zeta = 1$ we found that the zeroth-order term was not populated by the SWPG ($a_0 = 0$), however from the above equation we can clearly see that the non-π phase step has introduced a population in the zeroth-order. The percent of the total input energy that is placed into the zeroth order is $|a_0(\zeta)|^2 = \cos^2(\zeta\pi/2)$. Furthermore, from equation 3.20 we can also calculate the other orders for $\zeta \neq 1$,

$$\begin{aligned}
a_n(x_0, \zeta) &= \frac{1}{d} \int_{-\frac{d}{2}}^{\frac{d}{2}} e^{i\zeta\phi(\tilde{x},x_0)} e^{-in\frac{2\pi\tilde{x}}{d}} \, \mathrm{d}\tilde{x} \\
&= \frac{1}{d} \left[\int_{-\frac{d}{2}}^{-\frac{d}{4}+x_0} e^{i\zeta\pi-in\frac{2\pi\tilde{x}}{d}} \, \mathrm{d}\tilde{x} + \int_{-\frac{d}{4}+x_0}^{\frac{d}{4}+x_0} e^{-in\frac{2\pi\tilde{x}}{d}} \, \mathrm{d}\tilde{x} + \int_{\frac{d}{4}+x_0}^{\frac{d}{2}} e^{i\zeta\pi-in\frac{2\pi\tilde{x}}{d}} \, \mathrm{d}\tilde{x} \right] \\
a_n(x_0, \zeta) &= \mathrm{sinc}\left(\frac{n\pi}{2}\right) \sin\left(\frac{\pi}{2}\zeta\right) e^{i\frac{\pi}{2}(\zeta-1)} e^{-in\frac{2\pi x_0}{d}} \\
a_n(x_0, \zeta) &= a_n(x_0) \sin\left(\frac{\pi}{2}\zeta\right) e^{i\frac{\pi}{2}(\zeta-1)}.
\end{aligned} \tag{3.28}$$

From this equation, we can see that the non-π phase step has not populated the even diffraction orders, but the odd orders have an overall phase shift and are reduced in amplitude by a factor of $\sin(\zeta\pi/2)$. This should be expected because we saw from equation 3.27 that the

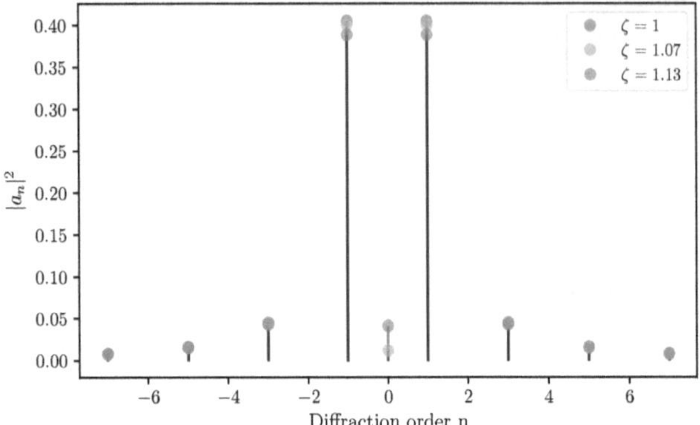

Figure 3.6: Square complex modulus of the Fourier coefficients a_n of the $0 - \zeta\pi$ SWPG for several values of ζ. The square complex modulus is proportional to the energy put into each diffraction order. As can be seen from figure, the ± 1 orders have the most energy put into them even for $\zeta \neq 1$. All non-zero even orders have zero energy.

zeroth-order was populated by a fractional energy of $|\cos(\zeta\pi/2)|^2$. Since this is a lossless system ($\sum_n |a_n|^2 = 1$), the energy that is populating the zeroth-order is coming from all of the odd orders that were populated. This redistribution of energy by $\zeta \neq 1$ is shown in 3.6.

From equations 3.27 and 3.28, we now have a notion of how the SWPG is behaving across the bandwidth of our femtosecond pulses. In particular, so long as ζ is close to 1, then the ± 1 orders are still the most intense, and as the grating offset x_0 is varied the phase difference between the ± 1 orders remains $\Delta\phi_{\pm 1} = 2\left(\frac{2\pi x_0}{d}\right)$ even though the overall spectral phase is modified by a factor of $e^{i\zeta\pi/2}$. To set a scale for what ζ close to 1 means, consider the case when the zeroth-order is equal in amplitude to the the ± 3 orders, $|a_0(x_0, \zeta)| = |a_{\pm 3}(x_0, \zeta)|$. In this case, $|\xi - 1| = |\frac{2}{\pi}\tan^{-1}(3\pi/2) - 1| \approx 0.13$. Therefore, it is reasonable to state the $0 - \pi$ SWPG maintains its phase control duplication properties for $|\zeta - 1| < 0.13$.

3.2.3 Square-wave phase grating design for high-harmonic generation

With the theory behind the SWPG well established, the specific grating parameters that were chosen with HHG in mind will be discussed in this section. The laser source that will be considered is the output of a HE-TOPAS optical parametric amplifier manufactured by Light Conversion. The TOPAS is pumped by a Spitfire ACE Ti:Sapphire system from Spectra-Physics. The Spitfire ACE system is capable of producing 12 mJ, 60 fs (20 nm

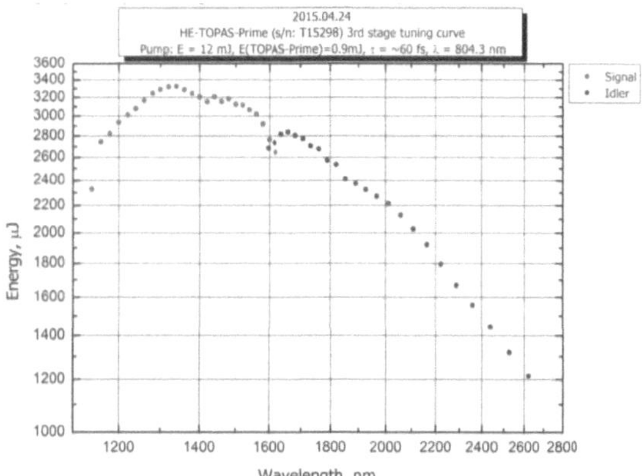

Figure 3.7: Pulse energy output of the TOPAS used in the experiments described within this chapter. The optimal output energy of the TOPAS can be seen to be around 1350 nm. This is the wavelength that was selected as the design wavelength for the SWPG.

FWHM bandwidth) pulses at 1 kHz. Using this system, the TOPAS is able to generate up to a combined 6 mJ of signal and idler. The signal wavelength range is from 1200 nm to 1600 nm, and within this range the TOPAS can output a nominally 70 fs pulse of up to 3 mJ with a tuneable central wavelength. A design wavelength of 1350 nm was chosen for the SWPG because the TOPAS performance is optimal around this wavelength (see figure 3.7).

Once the design wavelength for the phase grating is selected, then the physical size, L, of the step is determined by dispersion of the material selected from the relationship $\phi = \pi = 2\pi n L/\lambda$. For our phase gratings, Corning HPFS 7980 was used, and with this material $L \approx 0.47 \mu m$. The refractive index and the corresponding ζ parameter is shown in figure 3.8. The limitation that $\zeta = 1$ for only the design wavelength can be relaxed somewhat by introducing a nonzero angle of incidence between the incoming beam and the SWPG to effectively increase the optical path length of the step. If this angle is θ, then the ζ parameter can be written as

$$\zeta(\lambda, \theta) = \sec\theta\left(\frac{n(\lambda)\lambda_0}{\lambda n_0}\right) \tag{3.29}$$

where λ_0 is the design wavelength and n_0 is the refractive index at the design wavelength. This factor is shown in figure 3.8. From this figure, it is clear that even though the SWPG is designed for a specific wavelength it can be used over a broad range of wavelengths that

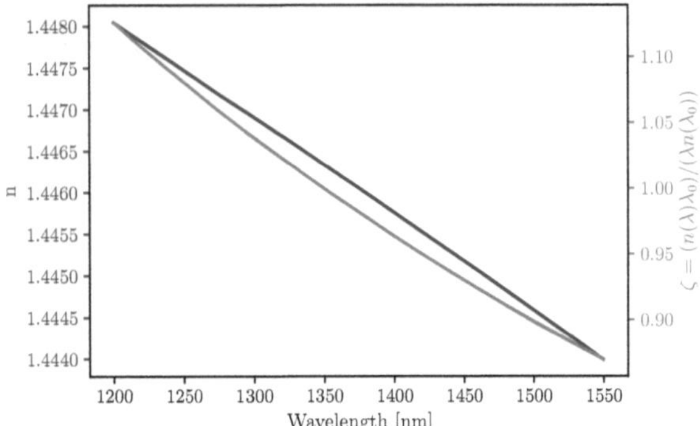

Figure 3.8: Refractive index (blue curve and axis) and the ζ parameter (red curve and axis). For a 70 nm bandwidth pulse centered at 1350 nm, $\zeta - 1$ varies from -0.026 to 0.027 assuming normal incidence.

are longer than the design wavelength. Of course, at higher angles of incidence propagation effects might become non-negligible, and those effects are neglected here.

The final remaining design parameter that must be considered is the choice of grating period. The choice of period is critical for the performance of the SWPG, and must be chosen with care. To get insight into how to select the correct period, we will reintroduce the β parameter from equation 3.11. For the specific situation we are considering, the β parameter is

$$\beta = \frac{2\pi\sigma D}{\lambda f} = \frac{2\pi\sigma\left(\frac{\lambda f}{d}\right)}{\lambda f}$$
$$= 2\pi\left(\frac{\sigma}{d}\right) \tag{3.30}$$

where σ is the input beam radius and $D = \tilde{x}_1 = \lambda f/d$ is that characteristic length scale at the focal plane because it represents the separation between the different diffraction orders in the focal plane. The beam profile at the focus can be calculated for various parameters of β, and is shown in figure 3.10. In the limit as $\beta \to 0$, the condition $d \gg \sigma$ must hold, and this condition implies that the source separation is approaching the waist radius of each diffraction order $\tilde{\sigma}$. Once the separation between orders becomes comparable to the waist ($\tilde{\sigma} \approx \tilde{x}_1$), then each diffraction order will strongly interfere with it's neighboring orders. The effect of this interference is that our sources can no longer be thought of as independent beams, and as the grating offset x_0 is varied there will be a strong modulation of both the amplitude and phase of each of diffraction order. This can be seen by using equation 3.22

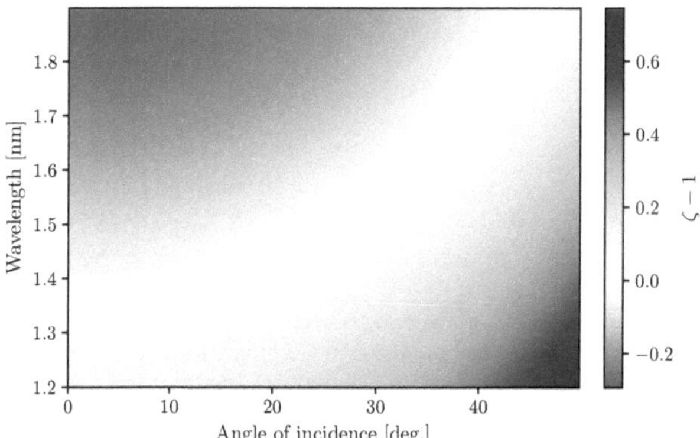

Figure 3.9: Non-π phase step parameter $\zeta(\lambda, \theta)$ calculated for a range of relevant wavelengths and incident angles. The ability to effectively tune the optical path length of the step enables the SWPG to be used for a much larger range of wavelengths that are longer than the design wavelength.

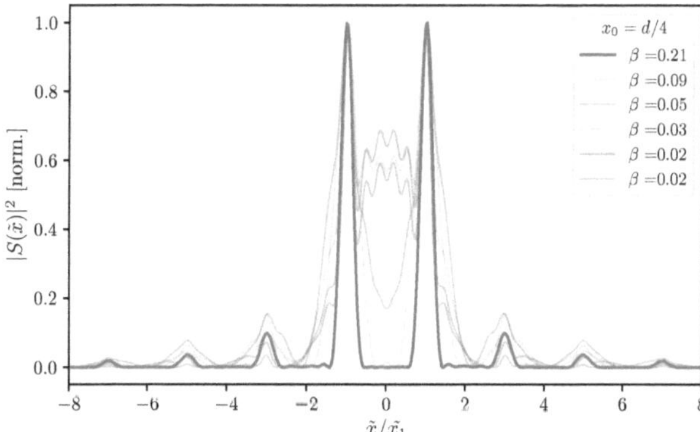

Figure 3.10: Calculation of intensity at the focal plane of the SWPG for various parameters of β. Calculation is done by numerically propagating the measured beam profile shown in figure 3.5. β is varied by adjusting the radius of the input beam profile. As $\beta \to 0$, one can see that the performance of the SWPG deteriorates and no longer produces well separated sources.

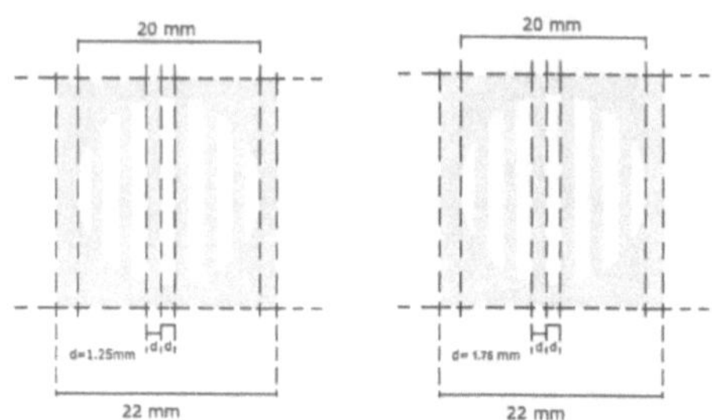

Figure 3.11: Schematic of the two SWPG which were purchased from Silios. They are constructed by etching the phase step in Corning HPFS 7980 fused-silica.

to write the intensity at the focal plane

$$|\tilde{S}(\tilde{x}, \phi_1)|^2 = \sum_{n=-\infty}^{\infty} \sum_{n'=-\infty}^{\infty} a_n \tilde{E}(\tilde{x} - \tilde{x}_n) e^{-in\phi_1} a_{n'} \tilde{E}(\tilde{x} - \tilde{x}_{n'}) e^{-in'\phi_1}$$

$$= \sum_{q=-\infty}^{\infty} e^{-iq\phi_1} \sum_{n=-\infty}^{\infty} a_n \tilde{E}(\tilde{x} - \tilde{x}_n) a_{n-q} \tilde{E}(\tilde{x} - \tilde{x}_{n-q}) \qquad (3.31)$$

$$= \sum_{n=-\infty}^{\infty} \tilde{E}_{2n+1}^2(\tilde{x}) + 2 \sum_{q=1}^{\infty} \cos(2q\phi_1) \sum_{n=-\infty}^{\infty} \tilde{E}_{2n+1}(\tilde{x}) \tilde{E}_{2n-2q+1}(\tilde{x})$$

where $\tilde{E}_n(\tilde{x}) = |a_n| \tilde{E}(\tilde{x} - \tilde{x}_n)$. The second term demonstrates that as the grating offset x_0 is varied, the intensity of a diffraction order $2n + 1$ will be modulated by an oscillatory term $\cos(2q\phi_1)$. The amplitude of this oscillation is determined by the overlap of the diffraction orders. Thus, for a grating such that $d \gg \sigma$ the oscillations will be very large because the source separation will be comparable to the beam waist of each order.

While it has become clear that choosing a grating period such that $\sigma \gg d$ is the ideal case for the performance fo the SWPG as a beam duplicator, there is another consideration that must be made for our specific application. In particular, we would like to generate high-harmonics from the ± 1 orders, and it order to do that we must send both sources through a gas medium generated by a gas jet in vacuum. This becomes increasingly difficult to handle as the source separation becomes large because the nozzle diameter must also increase, and the throughput of the nozzle increases quadratically with the diameter [53]. Since this needs to be done in vacuum, at a certain point the pumping requirements become untenable. So, in light of these considerations, the grating periods that were chosen were 2.5 mm and 3.5 mm. A schematic of the gratings are shown in figure 3.11. Using these gratings and a $f = 400$ mm lens at 1450 nm, we are able to achieve a source separation of 331 μm and 464

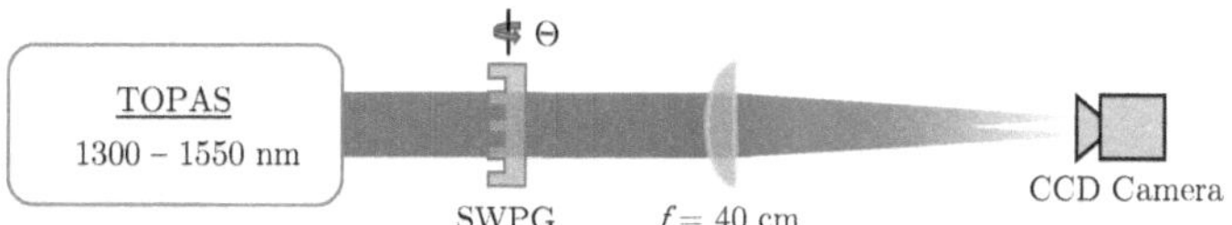

Figure 3.12: Schematic of the measurement used to verify the performance of the SWPG by measuring the beam profile at the focal plane using a CCD camera. The wavelength was varied from 1300 - 1550 nm, and the incidence angle θ was varied from 0-50°.

μm for the grating periods $d = 3.5$ mm and $d = 2.5$ mm. These parameters correspond to $\beta \approx 7$ for the 3.5 mm grating and $\beta \approx 10$ for the 2.5 mm grating for an input beam radius of $\sigma = 4$ mm.

To verify the expected behaviour of the SWPG as a duplicator, the intensity profile at the focal plane was measured for a variety of wavelengths and incidence angles for the SWPG with a grating period of $d = 3.5$ mm. The setup to perform this measurement is shown in figure 3.12. The wavelength was varied from 1300 - 1550 nm, and the incidence angle was varied from 0-50°. For each combination of wavelength and incidence angle an image was taken of the intensity profile. An example is shown in figure 3.13, and this corresponds to a wavelength of 1350 nm and an incidence angle of 0°. These parameters represent the design condition of the SWPG, and excellent agreement is seen between the measured intensity profile and the one calculated by numerically propagating the beam profile shown in figure 3.5 (a) for $\zeta = 1$.

This measurement also allows the performance of the SWPG outside of the design conditions to be investigated. An example of just such a case is shown in figure 3.14, and this corresponds to a wavelength of 1350 nm and an incidence angle of 30°. From equation 3.29, we expect that the non-π phase step parameter ζ will not be unity for this case, and this is confirmed by the agreement between the measured intensity profile and the numerical propagation for $\zeta = 1.19$. As can be seen in the intensity profile, the amplitude of the zero-order diffraction from the SWPG is non-zero due ζ not being unity, as was predicted by equation 3.27. The dependence of the zero-order amplitude on wavelength and incidence angle is shown in figure 3.15, and this compared to the expected dependence based upon equations 3.29 and 3.27. Reasonable agreement is shown between the measured and calculated zero-order amplitude.

These measurements of the intensity profile at the focal plane confirm the capability of the SWPG to generate two duplicates of an input IR pulse as the ± 1 diffraction orders. However, only the intensity profile is measured for a variety of input conditions, and this does

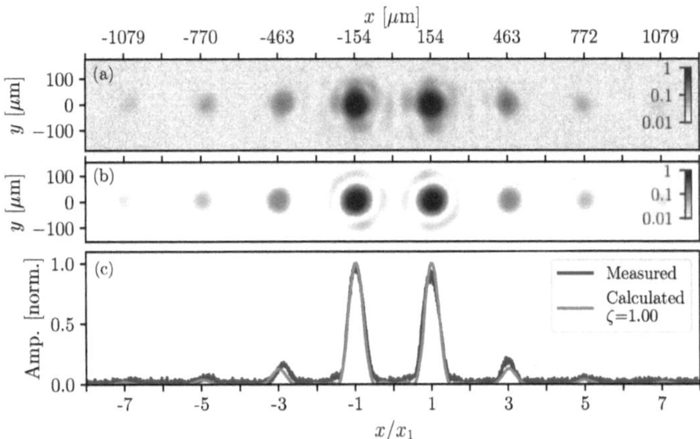

Figure 3.13: (a) Measured intensity profile at the focal plane of the SWPG for a wavelength of 1350 nm and an incidence angle of 0°. (b) Calculated intensity profile by numerical propagation for a wavelength of 1350 nm and $\zeta = 1$. (c) Line outs for $y = 0$ mm showing the different diffraction orders. Reasonable agreement is demonstrated between calculation and measurement.

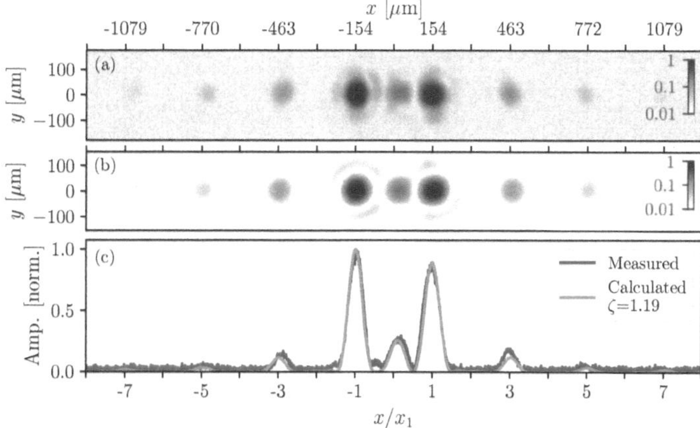

Figure 3.14: (a) Measured intensity profile at the focal plane of the SWPG for a wavelength of 1350 nm and an incidence angle of 30°. (b) Calculated intensity profile by numerical propagation for a wavelength of 1350 nm and $\zeta = 1.19$. (c) Line outs for $y = 0$ mm showing the different diffraction orders and a pronounced zero-order amplitude.

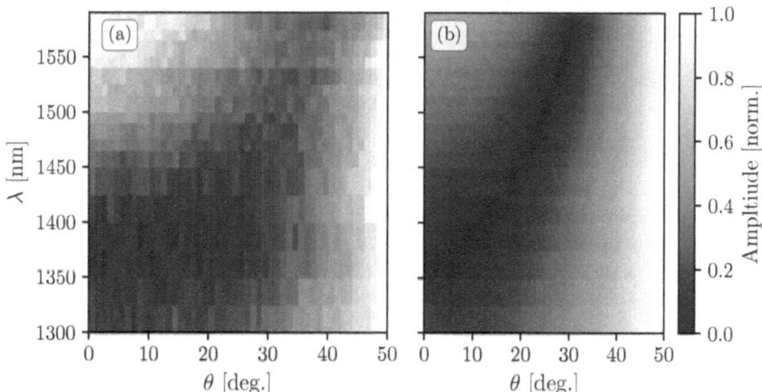

Figure 3.15: (a) Measured zero-order diffraction amplitude as a function of wavelength and incidence angle. (b) Zero-order amplitude calculated using equations 3.27 and 3.29.

not demonstrate the expected ability of the SWPG to be able to control the relative phase between the ±1 orders. To demonstrate this capability, high-harmonics will be generated from the ±1 orders, and the interference between the generated high-harmonics will reveal the relative phase control.

3.3 Two-source high-harmonic generation

To demonstrate that the $0 - \pi$ SWPG can be used as a femtosecond beam duplicator with relative phase control, we will generate harmonics from the ±1 diffraction orders. This experiment will be performed in the transient-absorption beamline (TABLe), and a schematic is show in figure 3.16. A $0 - \pi$ SWPG will be used to generate two intense lobes at the focal plane of a CaF_2 plano-convex lens with a focal length of 400 mm. A piezoelectric pulsed valve gas jet with a nozzle diameter of 500 μm is placed near the focus to deliver a gas medium in which high-harmonics are generated. The gas that will be used for generation will be argon. An image of the two sources is shown in figure 3.17. The laser which was used for this experiment is the output of an HE-TOPAS pumped by the Spitfire laser system. We will be working with a central wavelength of 1435 nm and a pulse energy of 2 mJ. The harmonics that are generated will pass through a 200 nm Al thin film to filter out the fundamental. In the energy ranges that we will be generating harmonics, Al will transmit harmonics in the energy range of 20 - 72 eV. After the metallic filter, the XUV will pass through a mirror with a hole in it and it will be refocused by an ellipsoidal mirror with a demagnification of 3. The XUV will then enter the spectrometer which consists of a Hitachi 1200 lines/mm variable line spaced (VLS) grating with a microchannel plate (MCP)/phosphor detector. The output of the phosphor is imaged by an Andor Neo 5.5 camera. The VLS grating focuses spectrally onto a flat-field, but in the transverse dimension it maintains the spatial profile. With this spectrometer we are able to simultaneously get spatial and spectral information about the incoming light.

The harmonics that are generated from this setup are shown in figure 3.18. Along the spatial dimension (labeled sensor position in the figure), one can immediately see a fringe pattern. These fringes are from the two sources that are generating harmonics. The intuitive way to understand the spatial frequency of these fringes is by thinking of them in terms of a Young's double slit. From this perspective, the wavelength dependence of the spatial frequencies present in the spatial profile of harmonic order q is

$$k_q = q\frac{2\pi\Delta x}{L\lambda} \propto q\hbar\omega \tag{3.32}$$

where L is the distance from the two sources to the detector and Δx is separation between the two sources. This linear dependence of the spatial frequency on photon energy is clearly seen in figure 3.18.

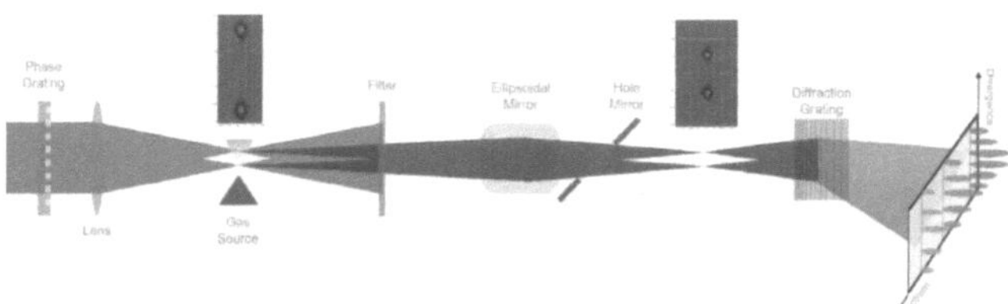

Figure 3.16: Schematic of the two-source HHG experiment performed in the TABLe. A $0 - \pi$ SWPG is used to generate two intense lobes at the focus of a lens. These lobes will generate XUV beams which will interfere in the far-field. An ellipsoidal mirror is used to refocus the XUV beams before going onto the spectrometer.

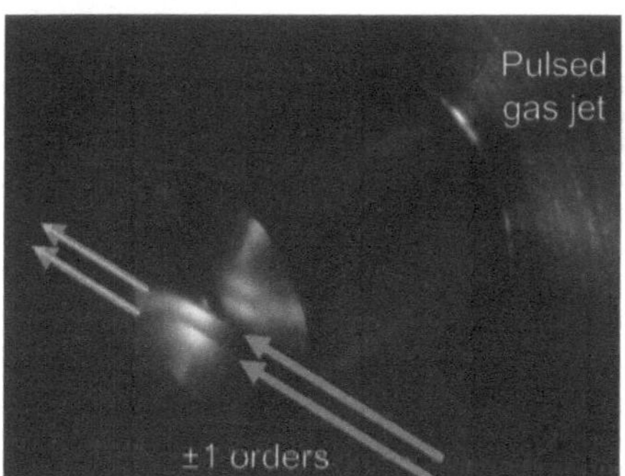

Figure 3.17: False color image of the two sources from the SWPG driving ionization in a gas medium delivered by the piezo gas jet shown in the image. Two sources follow the red arrows, and the XUV that is generated is shown by blue arrows.

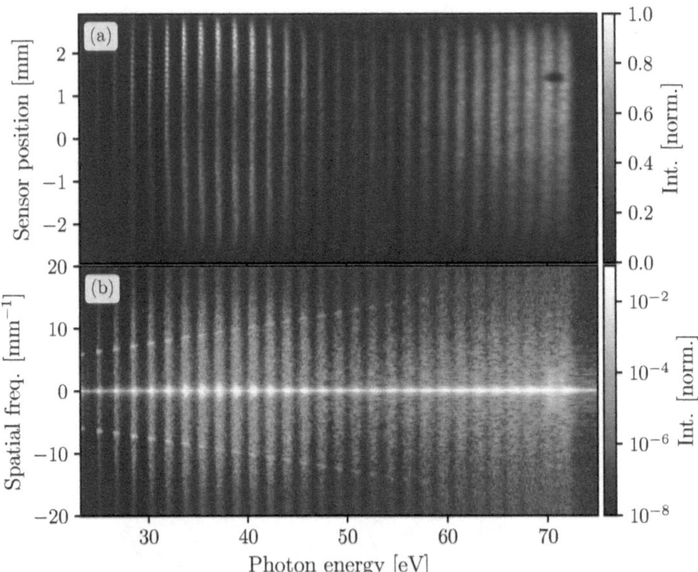

Figure 3.18: (a) Reference harmonic spectrum generated with a $0 - \pi$ SWPG. The fringes along the sensor position dimension are due to interference between the two XUV sources that are generated. The position of the fringe pattern is determined by the relative phase between the two sources. This relative phase can be controlled by the SWPG. (b) Power spectrum of the Fourier transform of the above image along the sensor position dimension. Clear peaks can be seen corresponding to the spatial frequency for each harmonic order. The linear dependence of the spatial frequency on photon energy is also seen.

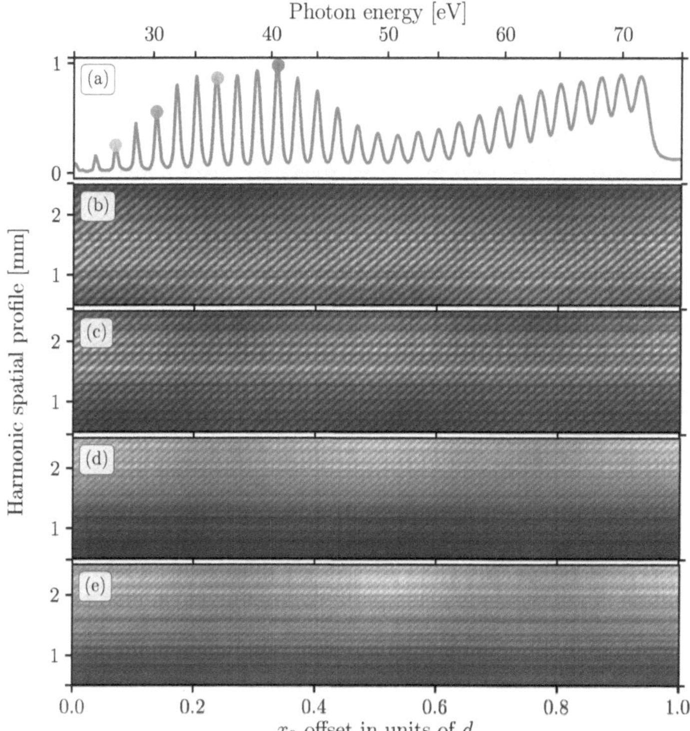

Figure 3.19: (a) Reference harmonic spectrum with dots showing the harmonic orders whose spatialgrams are plotted below. (b)-(e) Spatialgrams for harmonic orders 29, 33, 39, and 45. Tilted fringe pattern shows fringe shift due to phase shift induced by translating the phase grating. Modulations with a period of $d/2$ arise because of interference between the two generating sources.

The position of the fringe pattern in the spatial profile of the harmonics is determined by the relative phase between the two HHG sources. Therefore, any phase shift between the two sources will be imprinted upon the spatial profile of the harmonics as a fringe shift. We will utilize this sensitivity to demonstrate the capabilities of the SWPG. If we generate harmonics from the ± 1 orders of the SWPG, then as the grating offset x_0 is varied we would expect the fringe pattern for harmonic order q to shift by a factor of q multiplied by the phase shift between the two IR sources. Thus, for a translation of the grating by Δx_0 the q-th harmonic fringe pattern will shift by $4q\pi\Delta x_0/d$. If the grating is scanned through it's full period of d, then the phase difference between the two harmonic beams of order q will span a range of phases up to $4q\pi$. Due to the high non-linearity of HHG, this technique is very sensitive to small shifts in phase between the two beams, and it is precisely this sensitivity that will be leveraged in later experiments to extract more information in transient absorption experiments. It is important to be clear that we are not looking to measure the absolute phase difference between the two sources. Instead, we are primarily interested in our ability to measure a very small phase difference between the two sources introduced by the SWPG.

The measured effect of translating the grating is shown in figure 3.19. Four harmonic orders have been selected and their spatial profile have been plotted versus grating offset position x_0. These types of figures will be referred to as a spatialgram. Each spatialgram exhibits a tilted fringe pattern that corresponds to the fringe shift induced by a phase shift between the two XUV sources. As expected, the higher order harmonics have a higher frequency fringe pattern because of their shorter wavelength. If one counts the number of fringes over the full grating period scan, then one would find $2q$ fringes for harmonic order q. This provides a direct measure of the harmonic order q and is used in a calibration scheme for the spectrometer. The ability to measure the harmonic order q from the spatialgram verifies that the SWPG is able to control the phase difference between the two IR sources with a precision of a few mrad.

An additional feature of the spatialgrams in figure 3.19 is a slower modulation that has a period of $d/2$. This effect is present for all harmonic orders, and appears in a very similar way. This slower modulation is due to the interference between the two IR sources that were introduced in equation 3.31. The modulations with a period of $d/2$ is due to interference between the ± 1 and ∓ 1 diffraction orders. There is also a modulation with a period of d that is present in the spatialgrams. This modulation is due to interference between the 0th order and the ± 1 orders. In general, the frequency of the modulations is related to the separation between the diffraction orders that are interfering. These modulations are a limiting factor in using the phase grating for Fourier transform spectroscopy in the XUV.

Another interesting result from these measurements is that the symmetry of the attosecond pulse train (APT) that is generated can be observed because we are, in effect,

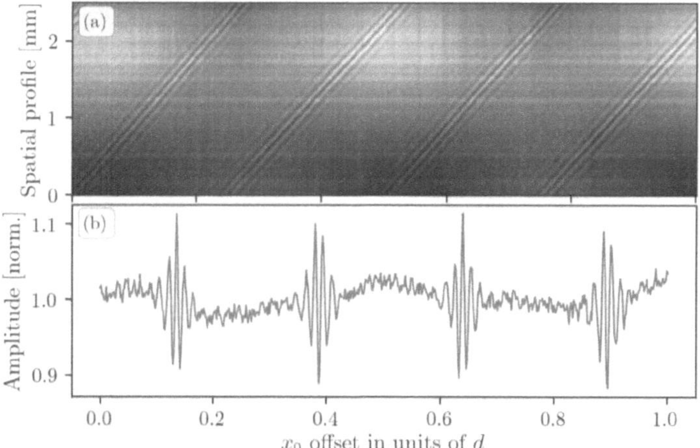

Figure 3.20: (a) Spatialgram of all harmonic orders combined. Diagonal stripes show the attosecond pulses that make up the APT. (b) Lineout from the full spatialgram.

measuring the interferometric autocorrelation of the XUV that is generated [54–57]. This is done either by looking at the zeroth-order diffraction off the VLS grating, or by integrating spectrally and looking at the combined spatial profile versus the grating offset. The later technique is used to generate figures 3.20 and 3.21. In figure 3.20, the pulses that make up the APT can be clearly seen. Since the phase grating is limited to a scan range of 4π between the two IR sources, we would expect to see only four pulses (one per half-cycle), and this is exactly what is observed. This can also be done in the case where the harmonics are generated using a two-color field consisting of the fundamental wavelength and it's second harmonic. In this instance, the symmetry of the field is broken, and one should expect to see an attosecond pulse once per cycle of the fundamental [58–60]. That exact case is shown in figure 3.21. In principle, by Fourier transforming these traces, one should be able to recover the harmonic spectrum with a spectral resolution of $\omega/2$.

3.4 Conclusion

In this chapter, the general concept of laser beam shaping was introduced, and the methods therein were applied to the specific problem of generating duplicates of a femtosecond IR pulse with precise control over their relative phase. The diffractive optical element that

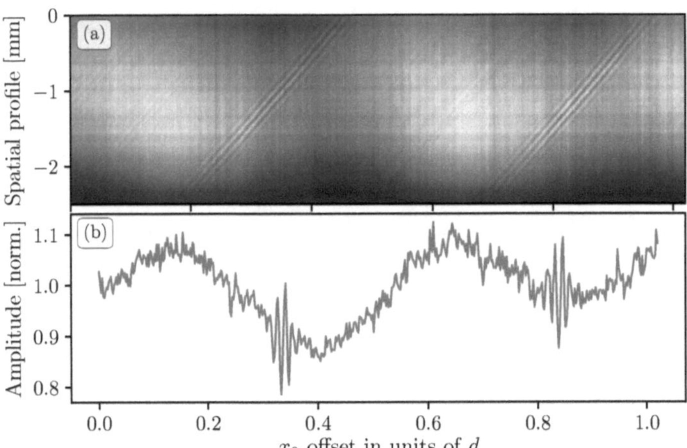

Figure 3.21: (a) Spatialgram of all harmonic orders combined. Diagonal stripes show the attosecond pulses that make up the APT. A second harmonic field was added to break the symmetry and generate even harmonics. (b) Lineout from the full spatialgram. Only two attosecond bursts are seen, which is expected from the asymmetric two-color generation field. The increased modulation with a period $d/2$ is due to the stronger interference between the two sources.

was shown to meet these demands is a $0 - \pi$ square-wave phase grating (SWPG). It's properties were discussed, and the final design parameters were chosen to optimize the SWPG for high-harmonic generation. The properties of the SWPG were demonstrated by generating high-harmonics and observing their corresponds fringe shifts. This experiment has also demonstrated the ability to measure small phase shifts between the two relative phase locked XUV beams. This property will be leveraged in experiments described in the following chapters.

Chapter 4
TWO-SOURCE FOURIER TRANSFORM SPECTROSCOPY

4.1 Introduction

In chapter 3, the $0 - \pi$ square-wave phase grating (SWPG) was introduced as a means of generating two intense duplicates of an input femtosecond IR pulse. An additional element of the SWPG is that it enables precise control over the relative phase between these two sources. When used to generate high harmonics, this scheme enables the generation of two XUV sources whose relative phase is well controlled by the SWPG, and any small phase shift between the two harmonic beams is imprinted upon their interference pattern as a fringe shift in the far-field. The idea is to now leverage this sensitivity to measure an induced phase shift between the two XUV sources. In the experiment described in this chapter, the phase shift will be induced by introducing a thin condensed matter sample into only one of the two XUV sources. Doing so enables us to extract both the real and imaginary part of the refractive index over a broad range of photon energies in the XUV for two different materials: germanium and silicon. This is done using two different methods to extract the refractive index from the interference pattern of the two XUV sources. One method involves extracting the change in fringe contrast and fringe shift directly, and the other method is an implementation of Fourier transform spectroscopy using the translation of the SWPG to control the phase between the two XUV sources.

4.2 Complex refractive index

The complex refractive index depends strongly on photon energy, and a cartoon of this is shown in figure 4.1. We are interested in the refractive index in the XUV energy region, and in this region there are many resonances that correspond to transitions of core-level electrons to unoccupied states near the Fermi level (for the case of a condensed matter system)[25, 61]. Complicated fine structure can emerge near these resonances that correspond to the local

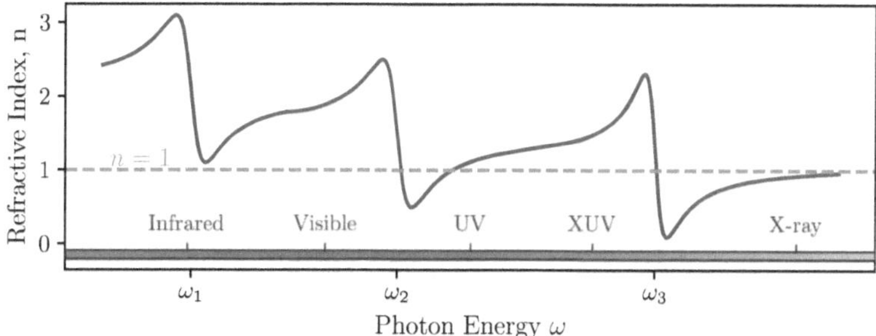

Figure 4.1: Schematic of the real part of the refractive index versus photon energy. Example resonances are shown in the IR, the visible/UV, and in the XUV/soft x-ray regimes. In general, the refractive index approaches 1 at higher photon energies.

electronic and geometric environment[25, 61]. Thus, the ability to measure both the real and imaginary parts of the complex refractive index can be important for many experiments using XUV light generated by HHG[62, 63].

In general, the complex refractive index can be written as

$$n(\omega) = 1 - \left(\frac{n_a r_e \lambda^2}{2\pi}\right) \left[f_1(\omega) - i f_2(\omega)\right] \tag{4.1}$$

where f_1 and f_2 are the atomic scattering factors, n_a is the number density, ω (λ) is the photon energy (wavelength), and

$$r_e = \frac{e^2}{4\pi\epsilon_0 m c^2} \tag{4.2}$$

is the classical electron radius [25]. By introducing the parameters β and δ, such that

$$\delta = \frac{n_a r_e \lambda^2}{2\pi} f_1(\omega)$$
$$\beta = \frac{n_a r_e \lambda^2}{2\pi} f_2(\omega), \tag{4.3}$$

then the refractive index n can be written as

$$n(\omega) = 1 - \delta + i\beta. \tag{4.4}$$

The values of both δ and β have been tabulated for elements up to uranium in the range of 10 eV to 30 keV[23], and their values are generally smaller than unity when far from resonance.

Now that we've established the form of the refractive index, we will consider the case of propagation through a dispersive medium and its effect on amplitude and phase of a wave [25]. The idea is to consider a plane wave of the form

$$\mathbf{E}(\mathbf{r}, t) = \mathbf{E}_0 e^{-i(\omega t - \mathbf{k} \cdot \mathbf{r})}, \tag{4.5}$$

and assume that the dispersion of the medium takes the form

$$\frac{\omega}{k} = \frac{c}{1 - \delta + i\beta}. \tag{4.6}$$

With these relationships, one can write the field in the propagation direction defined by $\mathbf{k} \cdot \mathbf{r} = kr$ as

$$E(r, t) = E_0 \left(e^{-i\omega(t - r/c)} \right) \left(e^{-i(2\pi\delta/\lambda)r} \right) \left(e^{-(2\pi\beta/\lambda)r} \right). \tag{4.7}$$

The first term in parentheses is the wave propagation, the second term is a phase shift proportional to δ that is induced by the dispersive medium, and the third term is a decay in amplitude that is proportional to β. From this relationship, it can be shown that the attenuation of the intensity is given by

$$\frac{I}{I_0} = e^{-(4\pi\beta/\lambda)r} = e^{-n_a \sigma_a r} \tag{4.8}$$

where I_0 is the initial intensity and $\sigma_a = 2r_e \lambda f_2(\omega)$ is the photoabsorption cross section. This relationship shows that by measuring the absorption of a material (a thin, free-standing film for these photon energies), one can easily extract the imaginary part of the refractive index.

The effect of the real part of the refractive index is to induce a phase shift in the propagating field, as can be seen from equation 4.7. After propagating through a material of thickness L, the induced phase shift is given by

$$\Delta\phi = \frac{2\pi\delta L}{\lambda}. \tag{4.9}$$

To experimentally access this phase shift, a technique that can be used is interferometry [64–66]. The idea is to create a Mach-Zehnder interferometer (see figure 4.2), and in one of the arms introduce a sample of thickness L. By measuring how the interference patterns shift when introducing the sample, then one can directly measure the phase shift induced by the sample. Additionally, by looking at how the fringe contrast changes, one can also get access to the attenuation caused by the sample. This means that both the real and imaginary parts of the refractive index can be probed simultaneously. This concept is precisely what will be used to extract the real and imaginary parts using the two XUV sources generated by a SWPG. In that case each source will act as one arm of a Mach-Zehnder interferometer.

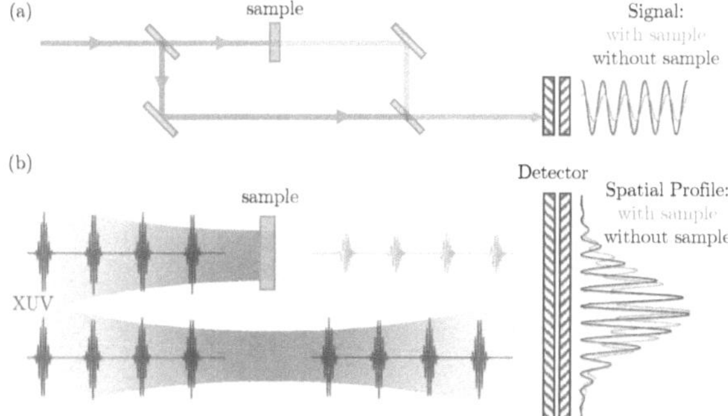

Figure 4.2: (a) Schematic of a Mach-Zehnder interferometer that is used to measure the phase shift induced by a sample placed in one of the arms of the interferometer. (b) For the experiments described in this chapter, the two XUV sources will act as the two arms of a Mach-Zehnder, and the sample of interest will only be introduced into one one the sources.

4.3 Measurement of the complex refractive index

4.3.1 Experimental setup

The experimental setup that will be used to demonstrate the ability to measure both the real and imaginary parts of the refractive index is very similar to the experimental setup presented in chapter 3. The TABLe is the experimental beamline that will be used, and the setup is shown in figure 4.3.

We use the output of the TOPAS at 1435 nm with a pulse energy of about 2 mJ and a pulse duration of around 60 fs. A $0 - \pi$ SWPG is with a grating period of 2.5 mm is used to generate two intense lobes at the focal plane of a 400 mm focal length CaF_2 plano-convex lens. At the focal plane, a gas medium is generated by a piezoelectric pulsed gas jet in which harmonics will be generated by the two sources. The generation gas that will be used is argon. The fundamental wavelength is then filtered out by an aluminum filter. The Al filter acts a high frequency band-pass with a band-pass region of 20-72 eV for the harmonic energies that are generated at this wavelength. The harmonics are then refocused into a target chamber by an ellipsoidal mirror with a demagnification of three. This entails that the source separation in the target chamber will be smaller by a factor of three, and the beam waist of each source will also be reduced by a factor of three. This is where a

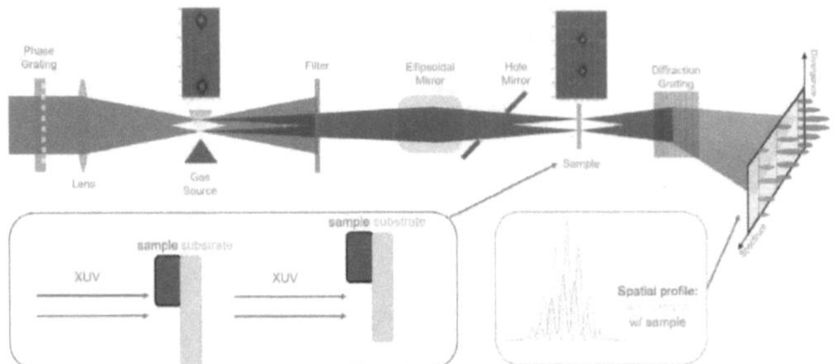

Figure 4.3: Schematic of the two-source HHG experiment performed in the TABLe. A $0 - \pi$ SWPG is used to generate two intense lobes at the focus of a lens. These lobes will generate XUV beams which will interfere in the far-field. An ellipsoidal mirror is used to refocus the XUV beams into a target chamber before going onto the spectrometer. A sample that is shaped like a step-function will be introduced at the focus of the XUV in the target chamber. The spatial profile of the various harmonic orders will be measured in the two cases shown. The fringe shift and fringe contrast changes allow for a simultaneous measurement of both parts of the refractive index of the sample.

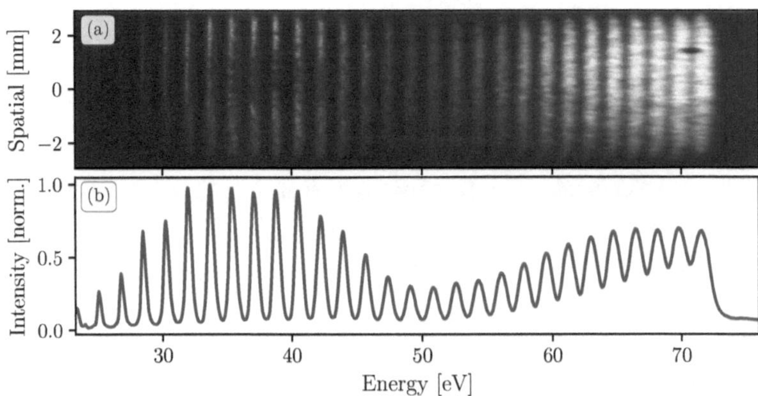

Figure 4.4: Reference image (a) and harmonic spectrum (b) that was used in this experiment. Fundamental wavelength is 1435 nm, and the harmonics are generated using an SWPG with a period of $d = 2.5$ mm.

sample will be introduced into only one of the two XUV sources. The sample in mounted on a motorized stage that allows for control of the position of the sample with respect to the XUV focus. After transmitting through the sample, the XUV will propagate to the spectrometer which allows for the spatial profile of each harmonic order to be measured. A reference image from the spectrometer and the corresponding harmonic spectrum used in this experiment is shown in figure 4.4.

In order to implement the scheme shown in figure 4.3, we need to introduce a sample into only one of the two XUV sources that are generated. As mentioned previously, the source separation in the target chamber will be a third of the separation in the generation chamber. For the SWPG and laser parameters that we used for this experiment, the source separation in the harmonic generation chamber is $\Delta x = 2\lambda f/d \approx 460\ \mu m$, and the corresponding separation will be $\Delta x_t = \Delta x/3 \approx 153\ \mu m$ in the target chamber. Therefore, the ideal sample has a cross sectional profile that is as close to a step function as possible, and the width of the step should be much less the separation between the two sources. In general, this can be accomplished using photolithography techniques to pattern a thin film of the desired profile on top of a free standing membrane substrate. For one of the samples that is used in this proof of principle experiment, we instead chose to start with a commercially available free standing membrane, and then break the membrane in such a way that it would have a sharp step-like cross sectional profile. The membrane that was chosen was a free standing 260 nm single crystal Si membrane on a 500 μm Si frame. These free standing membranes are manufactured by Norcada. The sample needs to be this thin because this experiment is done in transmission and XUV is strongly absorbed by most materials. A schematic of the sample that was used is shown in figure 4.5 (a). A second sample was also fabricated using e-beam deposition of Ge on top of a 30 nm SiN free-standing membrane.[4] For this sample, shown in figure 4.5 (b), a physical mask was used to cover half of the SiN membrane before deposition. Due to the delicate nature of these free-standing membranes, the physical mask could not touch the membrane without breaking it, and this necessitated a small gap between the physical mask and the membrane itself. The result of this gap is that there will be a more gradual change in thickness between the two halves of the sample (with and without Ge).

To be more precise about the cross sectional profile of the fabricated samples that were used, a form of profilometry was performed using the XUV beam as a probe of the thickness of the sample. The idea is simply an extension of the knife edge method that is used to measure beam profiles, see section 2.4.3 for the basics. In this case, the assumption that the knife is totally opaque is lifted and the transmission function of the sample is included.

---

[4]This was done by Yaguo Tang, a postdoc in the Agostini-DiMauro group.

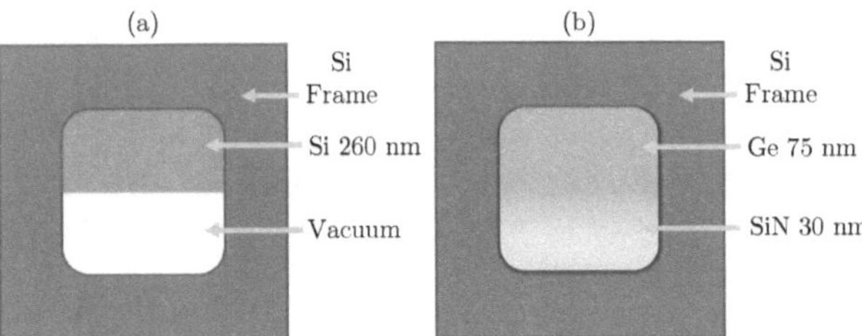

Figure 4.5: Schematic of the samples that were used in this experiment. (a) Free standing 260 nm Si membrane that has been broken in half. The way that the sample was cleaved in half ensures that the edge is sharper than the separation between the two sources. The sample was made by Norcada before it was broken. (b) Germanium deposited on a free standing 30 nm SiN membrane. E-beam deposition was performed with a physical mask to create a step-like profile.

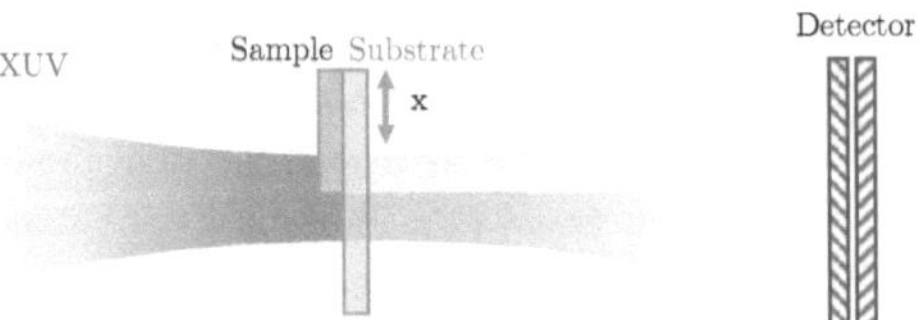

Figure 4.6: Schematic of knife edge technique used as a profilometry tool. A transmissive sample and substrate are translated through the focus of the XUV, and the detected harmonic amplitude as a function of position can be used to reconstruct the beam and transmission profiles.

The total transmitted power in this case now becomes

$$P(x,\omega) = \int_{\infty}^{\infty} \int_{\infty}^{\infty} T(x - x', \omega) I(x', y')\, \mathrm{d}x'\, \mathrm{d}y \tag{4.10}$$

where $T(x, E)$ is the transmission of the sample as a function of position and energy.[5] This essentially represents the spatial convolution of the transmission profile of the sample with the beam profile. Generally, in order for this to be useful as a profilometry tool, the beam size must be much smaller the feature sizes of the transmission profile. If this is not the case, then *a priori* knowledge of either the beam profile or the transmission profile is needed to reconstruct the other profile. As seen in section 2.4.3, the beam profile can be independently measured using part of the sample frame to determine the beam profile. If the beam width can be treated as vanishingly small relative to the transmission profile, then the total transmitted power becomes

$$P(x,\omega) = P_0(\omega)T(x,\omega). \tag{4.11}$$

To determine the thickness profile of the sample some assumptions are need. If one assumes knowledge of the absorption cross section as a function of energy, then Beer-Lambert's Law can be used directly to calculate the thickness, see equation 4.8. Additionally, if one assumes knowledge of the maximal thickness d_0 of the sample on top of the substrate, then the total transmitted power becomes

$$P(x,\omega) = P_0(\omega)T(x,\omega) = P_0(\omega)T_0(\omega)^{d(x)/d_0} \tag{4.12}$$

where T_0 is the transmission at the maximal thickness d_0. This allows for calculation of the thickness profile from the integrated harmonic signal, and the resulting relationship is

$$d(x) = d_0 \left(\frac{\log \overline{C}(x)}{\log \overline{T_0}} \right) \tag{4.13}$$

where $\overline{C}$ and $\overline{T_0}$ are the integrated average harmonic counts and maximal transmission.

This method is used to measure the profile of the samples that were used in this experiment. The results of these measurements are shown in figure 4.7. The integrated harmonic signal is shown in purple and shows a very sharp step for the Si sample whose width is limited by the harmonic beam waist of 6 μm, however the Ge sample shows a much broader transition region between the two regions that is approximately 200 μm in width. This was to be expected from the type of physical mask used to make the Ge sample, and it could be improved upon by implementing photolithography techniques. That being said, these samples are of sufficient quality to demonstrate the technique of measuring the complex refractive index using the SWPG.

---

[5]For the case of a opaque knife, $T = \Theta(x - x')$ and equation 2.11 is recovered.

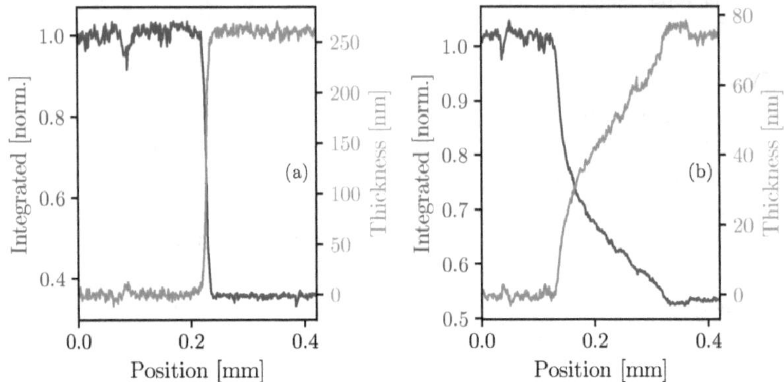

Figure 4.7: Integrated XUV signal (purple) as samples in shown in figure 4.5 are translated in the focal plane. Result is a profilometry measurement using XUV to characterize the thickness profile (blue) of the two samples with Si sample shown in (a) and the Ge sample in (b).

4.3.2 Sample Translation

To measure the complex refractive index of the samples shown in figure 4.5, we will look for a fringe shift and change in fringe contrast when only one of the sources is going through the sample and the other is going through either vacuum or the substrate. To see this, we translate the sample through the focal plane in the target chamber such that three distinct regimes will occur. The first is when both sources are going through vacuum/substrate, the second is when one source is going the sample and the other is going through vacuum/substrate, and the final regime is when both sources are going through the sample. Since this is a differential measurement, we would only expect to see a fringe shift for the second regime. The first and third regimes should show the same fringe pattern, and the only expected difference is the overall modification of the spectral amplitude of the harmonics due the absorption of the sample. This is shown in figure 4.8 (b) for the Si sample and figure 4.9 (b) for the Ge sample, where the spatial profile is shown for harmonic order 29 and 37 as the sample is translated through the focal plane.

From the spatial profile shown in figure 4.8 (b), the three expected regimes can clearly be seen. There is also additional spatial structure that is present in the transition between each of the three regimes. This is due to diffraction that is caused by one of the sources being partially blocked. From the spatial profile, the fringe shift can be extracted from the spatial frequency component that corresponds to this harmonic. The phase of that spatial frequency is plotted in 4.8 and shows that there is a phase shift between the two sources when only one of the sources is going through the sample. From this phase shift it is now possible to calculate the real part of the refractive index from the relationship

$$\delta = \frac{\lambda \Delta\phi}{2\pi \Delta d} \tag{4.14}$$

where Δd is the thickness difference and $\Delta\phi$ is the phase shift between the two sources. The phase shift that is observed in harmonic order 29 can be seen across the harmonic spectrum, and enables the real part of the refractive index to be extracted across a broad range of photon energies simultaneously. Similar behaviour is observed for the Ge sample in figure 4.9 (b), however the broad thickness profile of the sample causes a gradual transition between the three regimes as each sources transmits through varying thickness of Ge. That being said, the phase shift that is observed can still be used to measure the real part of the refractive index, just at an effective thickness given by the difference in thickness of Ge between each source.

In addition to the fringe shift that is show in figure 4.8 (b), there is also a change in fringe contrast that can be seen as the sample is translated through the two sources. In general, the fringe contrast can be defined as the relative difference of of the maximum and minimum values of an interference pattern, such that

$$V = \frac{I_{\mathrm{max}} - I_{\mathrm{min}}}{I_{\mathrm{max}} + I_{\mathrm{min}}} = \frac{I_{\mathrm{amp}}}{I_{\mathrm{mean}}} \tag{4.15}$$

is the fringe visibility or contrast. When considering the case of two interfering beams, this fringe contrast can be written as

$$V = \frac{2\sqrt{I_1 I_2}}{I_1 + I_2} |\gamma| \tag{4.16}$$

where I_1 and I_2 are the intensity of the two beams and γ is the complex coherence[6] [65–67]. The change in fringe contrast due to the sample translation is shown in figures 4.8 (c) and 4.9 (c). The contrast shows the same three distinct regimes that were seen in the fringe shift. Similarly to the fringe shift, it can be seen that there is only a change in fringe contrast when only one of the sources is going through the sample. From this contrast, it is possible to calculate the imaginary part of the refractive index. This can be done using the relationship

$$\beta = -\frac{\lambda}{2\pi \Delta d} \ln\left[\frac{V_0}{V}\left(1 - \sqrt{1 - \left(\frac{V}{V_0}\right)^2}\right)\right] \tag{4.17}$$

---

[6]The complex coherence is a function of harmonic order q and peak intensity I_0 of the fundamental field used to generate the harmonics, and it causes a decrease in fringe visibility for increasing q and I_0 [67].

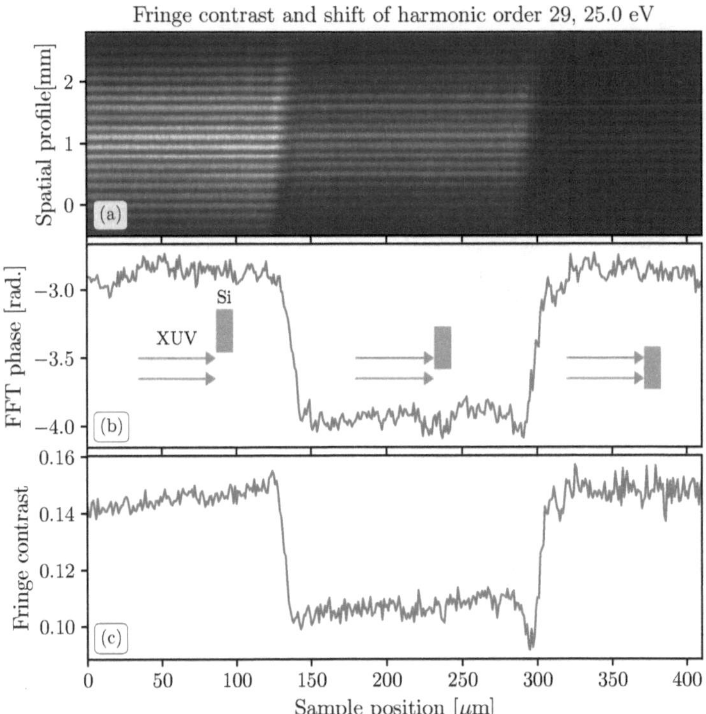

Figure 4.8: (a) Spatial profile of harmonic order 29 as the silicon sample is translated through the two sources. Three regimes are clear from the spatial profile, and they correspond to both sources going through vacuum, only one source going through the sample, and both sources going through the sample. A clear fringe shift can be seen between the second regime and the other two. Additional structure is seen at the transition between regimes, and this is due to diffraction caused by the sample partially blocking one of the sources. (b) Phase extracted from the spatial frequency corresponding to this harmonic order. The phase shift induced by the Si sample can be extracted from this phase shift. (c) Fringe contrast extracted from the spatial frequency corresponding to this harmonic order.

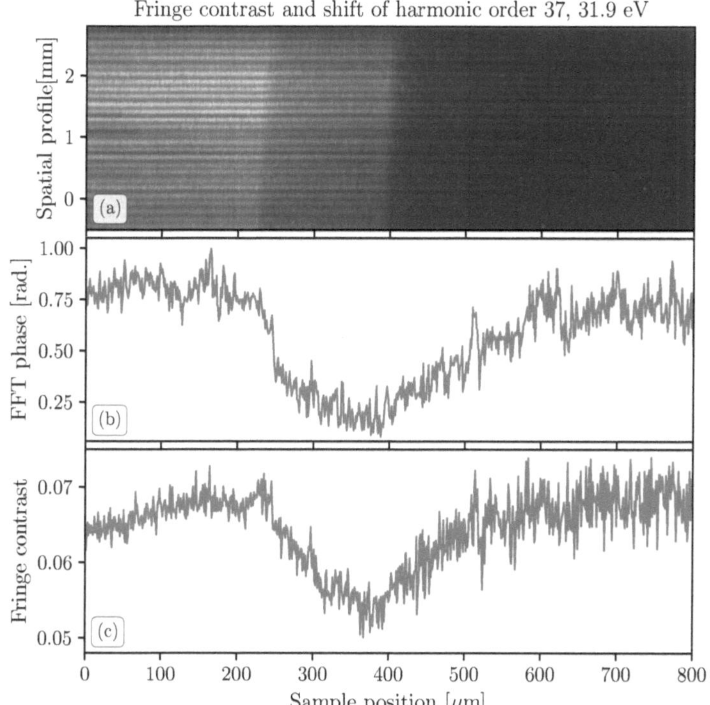

Figure 4.9: (a) Spatial profile of harmonic order 37 as the germanium sample is translated through the two sources. Three regimes can be seen, and they correspond to both sources going through vacuum, only one source going through the sample, and both sources going through the sample. The gradual transition between these cases is due to the sample profile shown in figure 4.7. (b) Phase extracted from the spatial frequency corresponding to this harmonic order. The phase shift induced by the Ge sample can be extracted from this phase shift. (c) Fringe contrast extracted from the spatial frequency corresponding to this harmonic order.

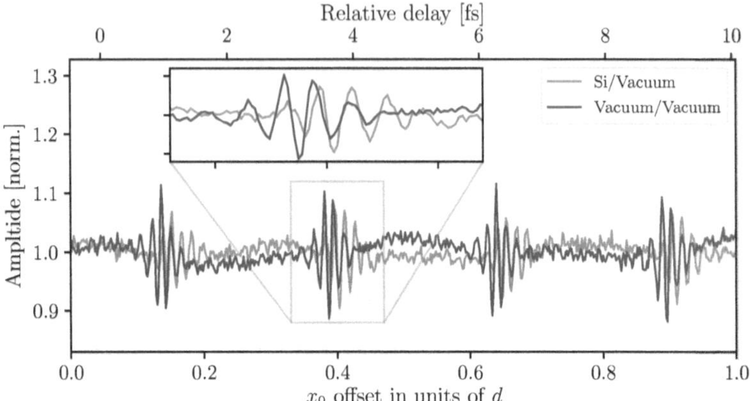

Figure 4.10: Interferogram of all harmonic orders that is extracted from the combined spatialgram with and without Si sample in one source. The shift between the two cases is 162 as in relative delay.

where V_0 is the contrast without the sample and V is the contrast with the sample present in one of the sources [65].

Therefore, the complex refractive index can be calculated from the apparent fringe shift and change in fringe contrast as the sample is translated through the two XUV sources. However, this can be improved upon by taking advantage of the unique capability of the SWPG that is used to generate the two sources, and that capability is the ability to control the relative phase between the two sources. This enables one to measure the complex refractive index using through a Fourier transform spectroscopy technique, and it enables a more accurate measure of the fringe shit and change in fringe contrast induced by the sample. These methods will be discussed in the following section.

4.3.3 SWPG Translation

An additional parameter that can be uniquely controlled by using the SWPG is the relative phase between the two XUV sources that are generated, as was established in chapter 3. To leverage this capability, the position of the SWPG can be translated through the beam profile to linearly vary the relative phase between the two sources at two different positions along the thickness profile of the samples. The obvious candidates for these positions along the thickness profile are: (1) only one source transmits through the sample and the other

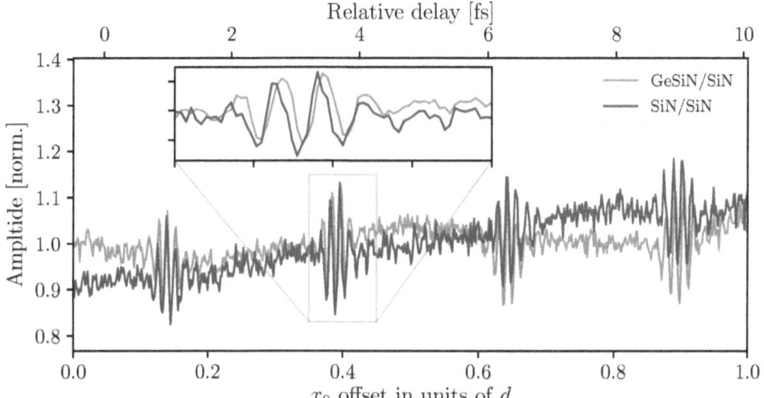

Figure 4.11: Spatialgram of combined harmonic orders with and without Ge sample in one source. The shift between the two cases is 16 as in relative delay.

transmits through the substrate/vacuum and (2) both sources transmitting through the substrate/vacuum. Intuitively, one would expect to see an additional phase shift between the two sources that is given by equation 4.9. To see how this phase shift manifests itself, the interferogram of all harmonic orders that is extracted from the combined spatialgram is shown in figure 4.10 and figure 4.11 for the two positions in the Si and Ge samples, respectively. For the case where both sources are going through the vacuum/substrate, this represents the inteferometric autocorrelation of the generated XUV. For the case where only one source is going through the sample, the combined spatialgram represents a cross-correlation between the reference source and it phase- and amplitude- modulated copy. Specifically, the detected signal is given by

$$S(x_0, \omega) = |A_{\text{sample}}|^2 + |A_{\text{ref}}|^2 + A_{\text{sample}} A_{\text{ref}} \exp\left(i\left(q\frac{4\pi x_0}{d} + \Delta\Phi(\omega)\right)\right) + c.c. \quad (4.18)$$

where $q = \omega/\omega_0$ is the effective harmonic order, A_{sample} and A_{ref} are the amplitudes of the XUV sources going through the sample and reference substrate/vacuum, $\Delta\Phi(\omega)$ is a phase shift between the two sources that includes the phase shift induced by the sample, and $q4\pi x_0/d$ is the phase shift introduced by the SWPG between the two sources [68]. In both figure 4.10 and 4.11, the structure of the APT can clearly be seen with an attosecond burst every half cycle of the fundamental, and, of more relevance, there is a clear phase shift that can be observed in the spatialgram between the two positions. To get a sense for the

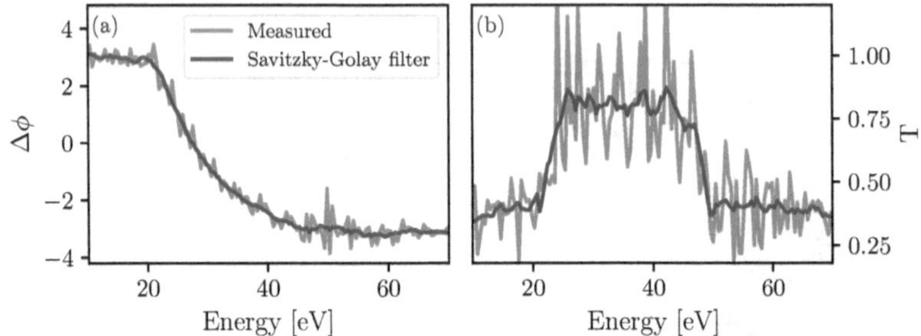

Figure 4.12: Spectral phase shift extracted from interferograms shown in figure 4.10. Phase shift is induced by the Si sample. Smoothed signal is shown using a Savitzky–Golay filter, and this is used in further calculations to extract the refractive index.

scale of the phase shift seen, the phase imparted by the SWPG between the two sources as a function of grating offset x_0 can be interpreted as a relative delay Δt between each pulse in the APT when the envelope is broad enough to be neglected. This yields a relationship given by

$$\Delta t = \left(\frac{\lambda_0}{c}\right)\left(\frac{x_0}{d}\right) \tag{4.19}$$

where λ_0 is the fundamental wavelength used to generate the two XUV sources. This delay corresponding to x_0 is also shown in figures 4.10 and 4.11, and the apparent shift between the two cases is 162 as for the Si sample and 16 as for the Ge sample. The ability to measure such a small shift highlights the capabilities of the SWPG because its inherent stability as a single optic interferometer enables this type of measurement. Additionally, a step size of 1 μm in x_0 that is easily achieved with many motors yields a relative delay of 1.9 as for $\lambda_0 = 1435$ nm and $d = 2.5$ mm, so this level of precision is easily obtained.

From the interferograms shown in figures 4.10 and 4.11, it is possible extract the spectrally dependent phase shift that is imparted by the sample. This is possible because those inteferograms constitute a Fourier-transform spectroscopy (FTS) measurement in the XUV using two phase locked beams generated by a SWPG. FTS enables measurement of both the real and imaginary parts of the refractive index, however it is difficult to perform in the XUV because of technical requirements set by the available optics and because the intrinsically short wavelengths in this energy range require high interferometric stability and precise control over the relative delay between the interfering beams [57, 68–70]. Both of those difficulties are solved through the use of a SWPG to generate the two sources that are used in this measurement. To extract the induced phase, the signal given by 4.18 is

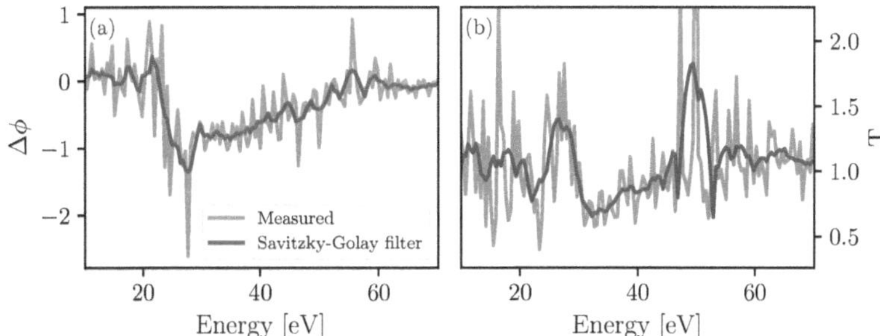

Figure 4.13: Spectral phase shift extracted from interferograms shown in figure 4.11. Phase shift is induced by the Ge sample. Smoothed signal is shown using a Savitzky–Golay filter, and this is used in further calculations to extract the refractive index.

Fourier transformed using the relative delay given by 4.19, and this yields

$$\tilde{S}(\omega) = A_{\text{sample}} A_{\text{ref}} \exp\left(i\Delta\Phi(\omega)\right). \tag{4.20}$$

In principle, the phase $\Delta\Phi$ includes geometric phase variations and phase variations from the HHG process [68]. To remove these contributions, a reference interferogram should be taken to isolate the phase contribution from the sample of interest, and this is precisely what was done by scanning the SWPG in the two positions mentioned previously. The sample induced phase $\Delta\phi$ is calculated from 4.20 for both positions through the relationship

$$\Delta\phi(\omega) = \arg\left[\tilde{S}_1(\omega)\tilde{S}_2^*(\omega)\right] \tag{4.21}$$

where $\tilde{S}_1(\omega)$ is the Fourier transformed interferogram measured at a position where only one source is on the sample and $\tilde{S}_2^*(\omega)$ the complex conjugate of the Fourier transformed interferogram measured at a position where neither source is on the sample of interest [68]. This analysis was performed to extract the phase shift induced by the Si and Ge samples shown in figure 4.5, and the resulting phase shift is shown in figure 4.12 (a) for the Si sample and figure 4.13 (a) for the Ge sample. Additionally, a smoothed signal is shown using a Savitzky-Golay filter in both figures. The phase induced by the Si sample shows a decreasing phase shift as the energy is increased with no apparent resonant structure, and this is expected for Si because there are no absorption edges corresponding to a core-level transition in this energy range [71]. The phase shift induced by the Ge sample, on the other hand, shows a pronounced resonance feature around 30 eV. This feature can be assigned to the $M_{4,5}$ absorption edge due to a core-level transition from the $3d$ states to states in the vicinity of the Fermi energy [71–76]. The signal is noisier in this case when compared to the Si sample because of the overall lower sample quality and the comparatively weaker harmonics in this energy range because of the absorption resonance and absorption from the substrate which does not contribute to the measured phase. This could be improved by fabricating a thinner sample with a more step-like thickness profile on a thinner membrane or by fabricating a free-standing membrane of Ge similar to the Si sample.

Beyond the sample induced phase shift, FTS can also give access to the imaginary part of the refractive index and thereby the energy dependent transmission of the sample. This can be calculated from the interferograms at the two positions along the sample using the relationship

$$
\begin{aligned}
T(\omega) &= \frac{A_{\text{sample}}(\omega)}{A_{\text{ref}}(\omega)} = \frac{A_{\text{sample}}(\omega)A_{\text{ref}}(\omega)}{A_{\text{ref}}(\omega)A_{\text{ref}}(\omega)} \\
&= \frac{|\tilde{S}_1(\omega)|}{|\tilde{S}_2(\omega)|}.
\end{aligned}
\tag{4.22}
$$

Applying this calculation to the interferograms measured for both samples is shown in figure 4.12 (b) and 4.13 (b) for Si and Ge, respectively. As can be seen from these figures, the transmission is noisier than the induced phase, however several features can still be observed in the energy range of 20-45 eV. In the case of Si, there is an increasing transmission that plateaus as the energy is increased. The drop is transmission past 45 eV is likely an artifact of the poor fringe contrast in the harmonics generated above 45 eV. This general behaviour agrees well with what is expected for Si [71]. For the case of Ge, the signal is even noisier than that of Si, however there is still a drop on transmission at 30 eV that corresponds to the $M_{4,5}$ absorption edge.

An additional way to extract the spectral phase and transmission induced by the sample is by examining the fringe shift and contrast as the grating position is varied in a similar manner as to what was done while the two samples were translated through the focus of the XUV, see figures 4.8 and 4.9. This can be done for each harmonic as x_0 is varied at the two positions along the sample thickness profile. The resulting fringe shift and contrast can then be averaged to give a measurement of the induced phase shift and absorption of the sample across the harmonic spectrum. This is shown for two harmonics in figure 4.14 for the Si sample. The phase ϕ corresponds to the phase of the spatial frequency of the harmonic, and it varies linearly with grating position x_0 via the relationship

$$
\phi = \left(q\frac{4\pi}{d} \right)x_0 + \phi_0.
\tag{4.23}
$$

As can be seen in the figure, the phase offset is different for the two positions where only one source is transmitting through the sample (solid line) and where neither sources are transmitting through (dashed line). The difference in phase between these two cases is precisely the sample induced phase that measured previously by the FTS method, and the

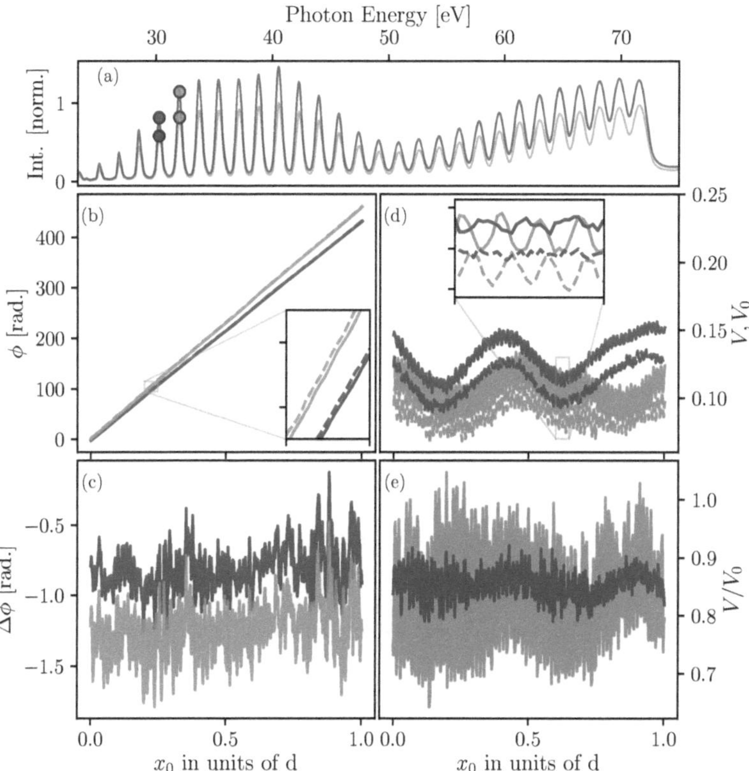

Figure 4.14: (a) Reference harmonic spectrum showing the two harmonics of interest. (b) Phase of the spatial frequency corresponding to the two harmonics of interest. (c) Difference in phase between the positions along the sample thickness profile. (d) Fringe contrast of each harmonic for the two positions along the sample. (e) Ratio of fringe contrast between the two positions. The sample in this case is silicon.

phase difference $\Delta\phi$ can be averaged over x_0 to give a more accurate phase shift at that harmonic energy.

In a similar way to the fringe shift, the fringe contrast can be measured as the phase grating is translated for the two sample positions. The resulting fringe contrast is shown in figure 4.14 (d) for the position where only one source transmits (V, dashed line) and where neither source does (V_0, solid line). As expected, the fringe contrast is greater for the case where neither source transmits through the sample because of the differential nature of this measurement. The modulations that are observed in the contrast as the SWPG is translated through the beam are due to interferences between the two sources in generation of the XUV beams. The leakage of one IR source into the other causes the slower modulations with a period of $d/2$, and the higher frequency modulations are due to interferences between the two sources at the harmonic energy. Regardless of their origin, the relevant quantity to extract the absorption induced by the sample is related to the ratio of contrasts V/V_0, see equation 4.17, and averaging V/V_0 over x_0 gives a more accurate change in fringe contrast.

Finally, we can now calculate both the real and imaginary part of the refractive index of Si and Ge over the range 25 - 60 eV by combining both the fringe shift and the change in fringe contrast and by using the FTS method.[7] The results are shown in figure 4.15 for Si and figure 4.16 for Ge. As can be seen in the figure 4.15, there is excellent agreement between the measured real and imaginary parts of the refractive index of Si when compared to the values that can be obtained from CXRO [71]. Both methods agree exceedingly well across most of the energy range of interest for the real part of the refractive index, however there are deviations between the FTS method (blue line) and the average fringe shift (purple dots) at energies above 45 eV. This is most likely due to the poor fringe contrast at energies above 45 eV, as seen in figure 4.4 (a). This deviation between the two methods is most pronounced in the imaginary part β shown in figure 4.15 (b). In principle, this measurement can be improved by generating harmonics with good contrast over the entire energy range of interest, and this can most easily be accomplished by using a grating with a larger period to reduce the overall spatial frequencies.

In this energy range, the Si refractive index is devoid of any resonance features because there no nearby absorption edges, so its measurement represents an easy benchmark of both methods of extracting the refractive index. However, what is typically of interest in most experiments is the refractive index at an absorption edge where there is strong variation in refractive index that is intrinsically related to the properties of the material in question. This is where the measurement of the refractive index of Ge, as shown in figure 4.16, comes in to play. As mentioned previously, the $M_{4,5}$ absorption edge in Ge is located at 30 eV, and its presence causes a large increase in absorption and a modulation of the phase around the absorption edge. The refractive index that has been tabulated by CXRO generally

---

[7]For the FTS method, the signal smoothed by a Savitzky-Golay filter is used. See figures 4.12 and 4.13.

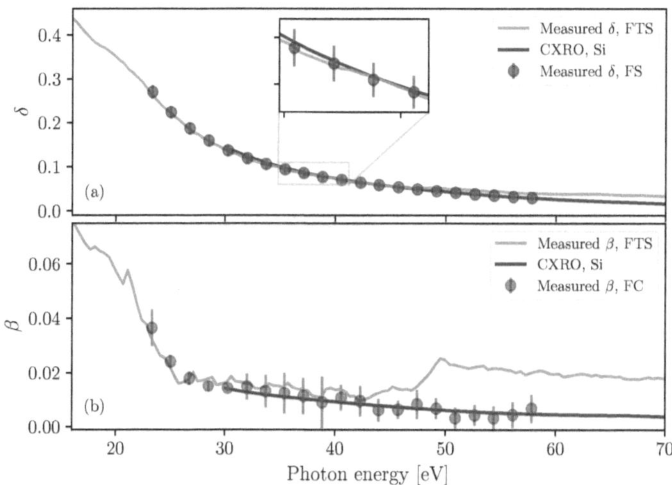

Figure 4.15: Real (a) and imaginary (b) part of the refractive index of Si measured with a SWPG using FTS (blue line) and averaging fringe contrast (FC) and fringe shift (FS) over x_0 (purple dots).

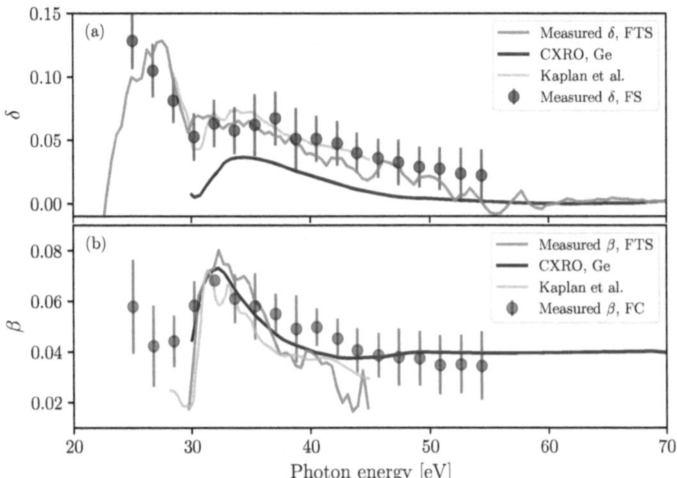

Figure 4.16: Real (a) and imaginary (b) part of the refractive index of Ge measured with a SWPG using FTS (blue line) and averaging fringe contrast (FC) and fringe shift (FS) over x_0 (purple dots).

is only accurate far from such resonance features. However, recent measurements of the refractive index were performed by Kaplan, *et. al.* [72], and this measurement serves as a good comparison to the two measurements methods presented herein. As can be seen from figure 4.16, there is generally excellent agreement between the two measurement methods and Kaplan, *et. al.* for the real part of the refractive index over the entire energy range of interest, however there are small deviations just above the edge between them. In this energy range just above the edge, the material properties of the sample strongly influence the refractive index, and further study of the sample properties could explain the deviations [25, 61, 72]. Similarly to the real part of the refractive index, the imaginary part, see figure 4.16 (b), shows good overall agreement from 30 - 45 eV in the vicinity of the absorption edge with small deviations that could possibly be explained with further analysis. However, for the imaginary part, the FTS fails outside the energy range of 30-45 eV because of a noisy signal. This could be improved both by fabricating a higher quality sample with a more suitable thickness profile and by optimizing the harmonics for better contrast over a larger energy. Regardless, this demonstrates the feasibility of using a SWPG to measure the complex refractive index over a large energy range in a single measurement even when a resonant feature is present.

4.4 Conclusion

In this chapter, the complex refractive index was introduced and a method to measure both the real and imaginary parts was proposed. The method relies on the use of a $0-\pi$ SWPG to generate two relative phase locked XUV sources whose interference acts as an inline Mach-Zehnder interferometer. By introducing a sample into one of the sources, the corresponding fringe shift and change in fringe contrast gives access to the real and imaginary parts of the refractive index. The intrinsic phase control offered by the SWPG allows for FTS to be performed by varying the SWPG position within the beam. Samples of Si and Ge were fabricated to test this method, and measuring their refractive index shows excellent agreement between our measured results and the literature. Thus, we have demonstrated the capability of using a SWPG to characterize the ground state complex refractive index of a condensed matter system. The next step is to extract the dynamic real and imaginary parts that are induced by dressing the sample with another IR field. This will be discussed in a later chapter.

Chapter 5
ATTOSECOND TRANSIENT-ABSORPTION SPECTROSCOPY

5.1 Introduction

In this chapter, the technique of attosecond transient-absorption spectroscopy (ATS) is used to study the dynamics of Fano resonances in a dressing field. The ATS technique was introduced in section 1.3 and will be further expounded upon in this chapter. The discussion herein begins with a theoretical review of Fano resonances, and this is followed by a review of recent theoretical efforts that have been used to explain the dynamics of Fano resonances in a dressing field. This theory is then applied[8] to the specific experiment that is performed in argon. This experiment is meant to serve as a prelude to the work that is done in chapter 6.

5.2 Autoionization resonances

One of the most extensively studied phenomenon using ATS has been autoionization of noble gas atoms in the time-domain [77–81]. Autoionization was first observed in 1935 by Beutler [82] by studying photoabsorption of noble gas atoms, and it manifested itself as sharp, asymmetric peaks in the absorption spectrum. These features were theoretically described by Fano in a seminal paper in 1961 [83, 84] as the result of interference between two pathways: direct ionization to the continuum and autoionization from a discrete state that is embedded in and coupled to the continuum. The theoretical framework that he developed can be treated as a more general formalism that describes interference between

---

[8]To be clear, all calculations within this chapter were performed by myself, however the underlying theory was developed by other research groups over many years. The relevant sources are cited throughout.

discrete and continuous pathways.[9] For this very reason, "Fano" resonances can be observed in a plethora of atomic, molecular, and condensed matter systems [88].

5.2.1 Autoionization in the frequency domain: Fano's original work

As noted above, Fano's theoretical explanation of the photoabsorption spectrum observed by Beutler in noble gas atoms is based on interference between two pathways. The relevant level diagram to describe this scenario is shown in figure 5.1, and specifically we will be considering the autoionization resonances in Ar because they will be used in the ATS experiments described in this chapter and in the following chapter. In this case, there is a bound state $|\psi_b\rangle$ (one of the $3s3p^6np$ states in Ar) that is embedded within a set of continuum states $|\psi_\varepsilon\rangle$. This entails that the energy of the bound state E_b is degenerate with the energetic spectrum of continuum states. The coupling between the bound state $|\psi_b\rangle$ and the continuum $|\psi_\varepsilon\rangle$ through the configuration interaction leads to decay of the electron from the bound state to the continuum. The following derivation of the photoabsorption cross section and phase follows closely from Fano's original paper and sources that have reproduced his original derivation [78, 84, 89].

The Hamiltonian describing this system can be written as

$$\hat{H} = \hat{H}_0 + \hat{V}, \tag{5.1}$$

where $\hat{H}_0$ is the zeroth order Hamiltonian and $\hat{V}$ is the correlation potential that describes the coupling between the discrete state $|\psi_b\rangle$ and the continuum state $|\psi_\varepsilon\rangle$. The solutions to the zeroth order Hamiltonian are the continuum and bound states, such that

$$\hat{H}_0 |\psi_b\rangle = E_b |\psi_b\rangle \tag{5.2}$$

$$\hat{H}_0 |\psi_\varepsilon\rangle = \varepsilon |\psi_\varepsilon\rangle \tag{5.3}$$

where the states $|\psi_b\rangle$ and $|\psi_\varepsilon\rangle$ are orthonormal. These two solutions to the zeroth order Hamiltonian are referred to as configurations, and the interaction between them is given by $\hat{V}$. The coupling strength between these two configurations is given by the off-diagonal matrix element V_ε, such that

$$\langle\psi_\varepsilon|\hat{H}|\psi_b\rangle = \langle\psi_\varepsilon|\hat{H}_0 + \hat{V}|\psi_b\rangle = \langle\psi_\varepsilon|\hat{V}|\psi_b\rangle = V_\varepsilon. \tag{5.4}$$

This configuration interaction matrix element V_ε depends upon the energy ε and is generally a smooth function of the continuous energy ε. Furthermore, the configuration interaction only couples different configurations and not within the same configuration. This means

---

[9]A very similar theory was independently developed by Feshbach in the context of nuclear physics, and these two theories have been unified by further theoretical work [85–87] .

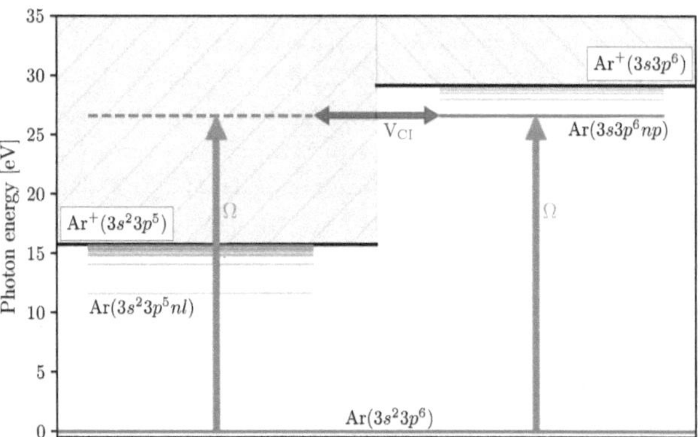

Figure 5.1: Level diagram of argon showing the effect of autoionization states on XUV photoabsorption. There are two possible pathways for ionization with a photon of energy Ω: (1) direct ionization to a continuum state (left side of figure) and (2) excitation to a bound state in the continuum (right side of figure). In case (2), there is coupling between the bound state and the continuum through the configuration interaction. This allows for the bound state to decay to the same continuum state as in case (1). This effect leads to interference between these two pathways.

that the diagonal matrix elements of $\hat{V}$ are zero,

$$\langle\psi_\varepsilon|\hat{V}|\psi_\varepsilon\rangle = 0 \tag{5.5}$$

$$\langle\psi_b|\hat{V}|\psi_b\rangle = 0. \tag{5.6}$$

Therefore, the diagonal matrix elements of the full Hamiltonian in equation 5.1 are given by

$$\langle\psi_\varepsilon|\hat{H}|\psi_\varepsilon\rangle = \varepsilon\delta(\varepsilon - \varepsilon') \tag{5.7}$$

$$\langle\psi_b|\hat{H}|\psi_b\rangle = E_b. \tag{5.8}$$

Armed with these states as a basis, we can now expand an eigenstate of the full Hamiltonian $\hat{H}$. This entails that the eigenstate $|\Psi_E\rangle$ of energy E, which is found by solving the equation

$$\hat{H}|\Psi_E\rangle = E|\Psi_E\rangle, \tag{5.9}$$

can be expanded in this complete basis, such that

$$|\Psi_E\rangle = a(E)|\psi_b\rangle + \int d\varepsilon' b(\varepsilon', E)|\psi_{\varepsilon'}\rangle. \tag{5.10}$$

The physical interpretation of this expansion is that an electron at energy E can originate from either the discrete state $|\psi_b\rangle$ or from the continuous state $|\psi_\varepsilon\rangle$. The contribution from $|\psi_\varepsilon\rangle$ is direct ionization, and the contribution from $|\psi_b\rangle$ is autoionization (i.e. decay from the bound state $|\psi_b\rangle$ to the continuum). The relative contributions of these two channels is given by the expansion coefficients $a(E)$ and $b(\varepsilon, E)$.

These expansion coefficients can be solved for using algebra described in full detail in Fano's paper [84]. The first step is to evaluate the relationship

$$\langle\Psi_E|\hat{H}|\Psi_E\rangle = E \tag{5.11}$$

using the expansion in eqn. 5.10. This results in a system of two equations with the unknown coefficients $a(E)$ and $b(\varepsilon, E)$. This system can be solved for analytical expressions of the expansion coefficients, and they are given by

$$a(E) = \frac{\sin\Delta(E)}{\pi V_E} \tag{5.12}$$

$$b(\varepsilon', E) = \frac{V_{\varepsilon'}}{E - \varepsilon'}a(E) - \delta(\varepsilon' - E)\cos\Delta(E) \tag{5.13}$$

where

$$\Delta(E) = -\arctan\left(\frac{\pi|V_E|^2}{E - E_b - F(E)}\right) \tag{5.14}$$

$$F(E) = \text{PV} \int \mathrm{d}\varepsilon' \frac{|V_{\varepsilon'}|^2}{E - \varepsilon'} \tag{5.15}$$

and PV is the Cauchy principal value. The term $F(E)$ is an energy-dependent shift of the bound state that depends upon the strength of the configuration interaction $|V_{\varepsilon'}|^2$. This shift can be either positive or negative, depending upon the sign of $\partial_{\varepsilon'}|V_{\varepsilon'}|^2$ at $\varepsilon' = E$, where $\partial_{\varepsilon'}$ is the partial derivative with respect to ε'. Thus, any change in $V_{\varepsilon'}$ by an external field will lead to a shift in the resonance position.

Substituting the coefficients in equations 5.12 and 5.13 into equation 5.10 yields

$$|\Psi_E\rangle = \frac{\sin \Delta(E)}{\pi V_E} |\psi_b\rangle + \frac{\sin \Delta(E)}{\pi V_E}\left(\text{PV} \int \mathrm{d}\varepsilon' \frac{V_{\varepsilon'}}{E - \varepsilon'}\right) |\psi_{\varepsilon}'\rangle - \cos \Delta(E) |\psi_E\rangle. \tag{5.16}$$

This can be further simplified by introducing a modified discrete state given by

$$|\Phi\rangle = |\psi_b\rangle + \text{PV} \int \mathrm{d}\varepsilon' \frac{V_{\varepsilon}'}{E - \varepsilon'} |\psi_{\varepsilon'}\rangle, \tag{5.17}$$

which allows us to express the eigenstate $|\Psi_E\rangle$ as

$$|\Psi_E\rangle = \frac{\sin \Delta(E)}{\pi V_E} |\Phi\rangle - \cos \Delta(E) |\psi_E\rangle. \tag{5.18}$$

Finally, the argument of equation 5.15 can be written in terms of an important parameter, the reduced energy given by

$$\epsilon = \frac{E - (E_b + F(E))}{\Gamma(E)/2} = \frac{E - E_\Phi}{\Gamma/2} \tag{5.19}$$

where

$$\Gamma(E) = 2\pi|V_E|^2 \approx \Gamma(E_b) = \Gamma. \tag{5.20}$$

The interpretation of the modified bound state $|\Phi\rangle$ is that the configuration interaction is mixing the original discrete state $|\psi_b\rangle$ and the continuum states $|\psi_{\varepsilon}'\rangle$. So, for an energetic window near $E = E_\Phi$, one can consider the resonance energy to be E_b and the resonance linewidth to be Γ. Since $\Gamma = 2\pi|V_E|^2$, the resonance linewidth and the natural lifetime h/Γ are directly related to the strength of the coupling between bound states and continuum states through the configuration interaction. Therefore, stronger (weaker) coupling would lead to faster (slower) decay from bound to continuum states, respectively. From this, it can be seen that an external field that is able to modify the strength of the configuration interaction, leading to a change in the linewidth and position of the resonance.

Now that the eigenstates of the Hamiltonian $\hat{H}$ have been expanded, we will turn our attention to the photoabsorption spectrum. In the original experiments done by Beutler, a sharp, asymmetric absorption profile was seen in the photoabsorption spectrum of noble gas atoms in the XUV [82]. From the expanded eigenstate given in equation 5.18, we can begin to see how this asymmetric absorption profile might arise. The coefficients in the expansion are proportional to sine and cosine functions of the reduced energy ϵ, and, since they are odd and even functions of ϵ, this will lead to constructive and destructive interference on either side of the resonance. It is precisely this effect that will give rise to the asymmetric absorption lineshape.

To derive the photoabsorption spectrum, we will consider a transition from the ground state of the atom $|g\rangle$ by a XUV photon of energy Ω. This can be described through the use of the dipole transition operator

$$\hat{D} = -e\hat{\mathbf{r}} \cdot \mathbf{E}_{XUV}(t) \tag{5.21}$$

where $\mathbf{E}_{XUV}(t)$ is the electric field of the XUV. Using this operator, the transition probability is given by the matrix element

$$\begin{aligned}
\langle \Psi_E | \hat{D} | g \rangle &= \frac{1}{\pi V_E^*} \sin \Delta(E) \, \langle \Phi | \hat{D} | g \rangle - \cos \Delta(E) \, \langle \psi_E | \hat{D} | g \rangle \\
&= \cos \Delta(E) \, \langle \psi_E | \hat{D} | g \rangle \left[\tan \Delta(E) \frac{1}{\pi V_E^*} \frac{\langle \Phi | \hat{D} | g \rangle}{\langle \psi_E | \hat{D} | g \rangle} - 1 \right].
\end{aligned} \tag{5.22}$$

At this point, we can now introduce the well-known and important q parameter, given by

$$q(E) = \frac{1}{\pi V_E^*} \frac{\langle \Phi | \hat{D} | g \rangle}{\langle \psi_E | \hat{D} | g \rangle} \approx q(E_b) = q. \tag{5.23}$$

The q parameter describes the asymmetry of the resonance, and it is related to the ratio of transitions to the modified bound state $|\Phi\rangle$ and the continuum states $|\psi_E\rangle$. Combining equations 5.22, 5.23, and 5.14, we arrive at

$$\langle \Psi_E | \hat{D} | g \rangle = -\cos \Delta(E) \, \langle \psi_E | \hat{D} | g \rangle \left[\frac{\pi |V_E|^2}{E - (E_b + F(E))} q + 1 \right] \tag{5.24}$$

$$= -\cos \Delta(E) \, \langle \psi_E | \hat{D} | g \rangle \left[\frac{\Gamma/2}{E - (E_b + F(E))} q + 1 \right], \tag{5.25}$$

and this can be further simplified using the reduced energy ϵ, which yields

$$\langle \Psi_E | \hat{D} | g \rangle = -\langle \psi_E | \hat{D} | g \rangle \cos \Delta(E) \left(\frac{q}{\epsilon} + 1 \right) \tag{5.26}$$

$$\langle \Psi_E | \hat{D} | g \rangle = \langle \psi_E | \hat{D} | g \rangle \frac{q + \epsilon}{\epsilon + i}. \tag{5.27}$$

Finally, using this relationship the ratio of transition probabilities can be calculated and

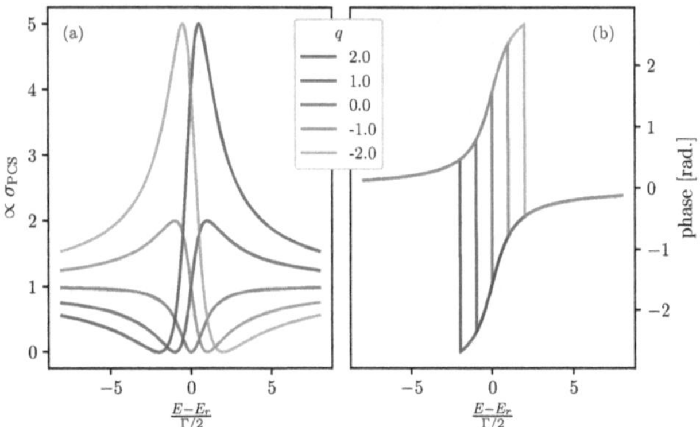

Figure 5.2: (a) Calculation of the photoabsorption cross section near a resonance for the listed q parameters. The change from symmetric to asymmetric profiles can be seen as the q parameter is varied. Maximum and minimum in cross section occus at $\epsilon = 1/q$ and $\epsilon = -q\Gamma^2/4$, respectively. (b) Calculation of the phase across the resonance for different q parameter. The π phase jump clearly depends on q, and it occurs when $\epsilon = -q$ and not at the resonance energy E_r. Calculations based on U. Fano's original work [84].

leads to the well known Fano lineshape,

$$\frac{|\langle\Psi_E|\hat{D}|g\rangle|^2}{|\langle\psi_E|\hat{D}|g\rangle|^2} = \frac{(q+\epsilon)^2}{\epsilon^2+1}. \tag{5.28}$$

This ratio is proportional to the photoabsorption cross section, and is plotted for various q parameters in figure 5.2. As can be seen, the lineshape's symmetry dramatically depends upon the q parameter, and the cross section even goes to zero at different energies, depending upon q. This is a direct consequence of the destructive interference from the configuration states, as was predicted earlier in the derivation. Additionally, the spectral phase of the Fano profile can also be extracted, given by

$$\theta(\epsilon) = \arg\left[\frac{q+\epsilon}{\epsilon+i}\right], \tag{5.29}$$

and is plotted in figure 5.2 (b). For increasing ϵ, the phase increases until $\epsilon = -q$ when there is a π phase jump, and thereafter the phase continues to increase until it asymptotically approaches its original value.

Experimentally the photoabsorption cross section is generally fit to the form

$$\sigma_{\mathrm{PCS}} = \sigma_a \frac{(q+\epsilon)^2}{\epsilon^2+1} + \sigma_{NR} \tag{5.30}$$

where σ_a scales the strength of the Fano profile and σ_{NR} is a non-resonant cross section that is included to account for absorption from other continuum states that might be present.

5.2.2 Autoionization in the time domain

Up until this point, Fano resonances have been discussed in a time-independent manner in the frequency domain. However, one would ideally like to describe these autoionizing resonances in the time domain, as that will lend itself to the experiments described herein. This description was primarily done by W.-C. Chu and C.D. Lin [90], and it has been used to interpret many ATS experiments [79, 80, 91, 92]. The basic assumption that is made in this treatment is that the XUV pulse that excites from the ground state to the Fano resonance is a Dirac-δ function in time, $\mathbf{E}_{XUV} = E_0 \delta(t)\mathbf{z}$. This impulsive excitation is generally a reasonable approximation for an attosecond XUV pulse that is shorter in duration than the typical lifetimes of these resonances, typically on the order of a few tens of femtoseconds. From this assumption, it is possible to analytically describe the dipole response of the system in the time domain.

The derivation of the dipole response follows naturally from the theory described in the previous section. Assuming that the XUV impulsively excites at $t = 0$ from the ground state $|g\rangle$ to the bound and continuum states $|\psi_b\rangle$ and $|\psi_\varepsilon\rangle$, the wave function for times $t > 0$ can be in the configuration states, such that

$$|\Psi(t)\rangle = e^{-i\varepsilon_g t} |g\rangle + c_b(t) |\psi_b\rangle + \int c_\varepsilon(t) |\psi_\varepsilon\rangle \, \mathrm{d}\varepsilon. \tag{5.31}$$

Since the configuration states $|\psi_b\rangle$ and $|\psi_\varepsilon\rangle$ are not eigenstates of the total Hamiltonian, the expansion coefficients c_b and c_ε are explicitly time dependent. The evolution of these coefficients is governed by the Time-Dependent Schrödinger Equation (TDSE)

$$i\frac{\partial}{\partial t} |\Psi(t)\rangle = \hat{H} |\Psi(t)\rangle, \tag{5.32}$$

and using this equation, the time dependence of the expansion coefficients can be expressed as the coupled differential equations given by

$$\begin{aligned}
\frac{\partial c_\varepsilon}{\partial t} &= -iV_\varepsilon c_b(t) - i\varepsilon c_\varepsilon(t) \\
\frac{\partial c_b}{\partial t} &= -i\varepsilon_r c_b(t) - iV_\varepsilon \int c_\varepsilon(t) \, \mathrm{d}\varepsilon.
\end{aligned} \tag{5.33}$$

Assuming the initial values $c_b^{(0)}$ and $c_\varepsilon^{(0)}$ are known, then the solutions to equations 5.33 are

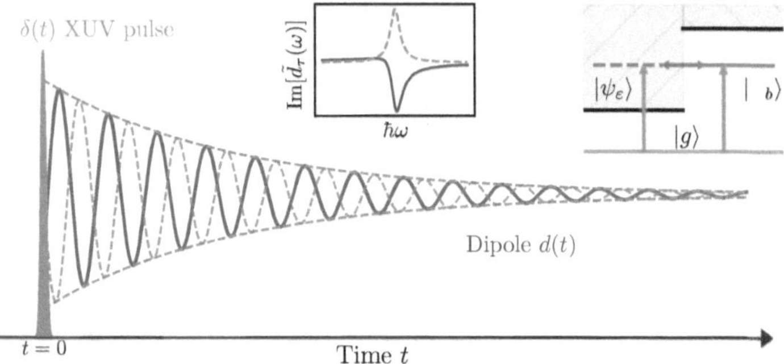

Figure 5.3: Illustration of the dipole moment after a $\delta(t)$ excitation pulse. Dipole is shown for a phase shift of $\varphi = -1.2\pi$ (purple curve) and $\varphi = 0$ (dashed gray curve), and the central inset shows the line shape for these two phase shifts.

given by

$$c_b(t) = c_b^{(0)}\left(1 - \frac{i}{q}\right)e^{-i\varepsilon_r}e^{-\frac{\Gamma}{2}t}$$

$$c_\varepsilon(t) = \frac{c_\varepsilon^{(0)}}{\epsilon + i}e^{-i\varepsilon_r t}\left[(q + \epsilon)e^{-i(\varepsilon - \varepsilon_r)t} - (q - i)e^{-\frac{\Gamma}{2}t}\right],$$

(5.34)

where, as in the previous section, $q = c_b^{(0)}/(\pi V c_\varepsilon^{(0)})$, $\Gamma = 2\pi|V|^2$, and $\epsilon = (\varepsilon - \varepsilon_r)/(\Gamma/2)$.

With the time dependent wave function now in hand, the induced dipole $d(t)$ can now be calculated for $t > 0$ by evaluating $d(t) = \langle\Psi(t)|z|\Psi(t)\rangle$. Using equations 5.31 and 5.34, this gives

$$d(t) = c_b(t)e^{-i\varepsilon_g t}\langle\psi_b|z|g\rangle^* + \int c_\varepsilon(t)e^{i\varepsilon_g t}\langle\psi_\varepsilon|z|g\rangle^* + c.c.$$

$$= c_b^{(0)}\langle\psi_b|z|g\rangle^* e^{-i\Omega_r t}\left[\left(1 - \frac{i}{q}\right)e^{-\frac{\Gamma}{2}t} + \frac{1}{(\pi V q)^2}\int\frac{(q + \epsilon)e^{-i\frac{\Gamma}{2}\epsilon t} - (q - i)e^{\frac{\Gamma}{2}t}}{\epsilon + i}\,d\varepsilon\right] + c.c.,$$

(5.35)

where $\Omega_r = E_r - E_g$ is the resonance photon energy. The complex conjugate term in equation 5.35 is a counter-rotating term, and by invoking the rotating wave approximation, it can be dropped. Equation 5.35 can further be evaluated to give

$$
\begin{aligned}
d(t) &= c_b^{(0)} \langle \psi_b|z|g \rangle^* e^{-i\Omega_r t} \frac{1}{q^2} \left[q(q-i)e^{-\frac{\Gamma}{2}t} + \frac{2}{\pi\Gamma} \int \frac{(q+\epsilon)e^{-i\frac{\Gamma}{2}\epsilon t} - (q-i)e^{\frac{\Gamma}{2}t}}{\epsilon+i} \, d\varepsilon \right] \\
&= c_b^{(0)} \langle \psi_b|z|g \rangle^* e^{-i\Omega_r t} \frac{1}{q^2} \left[q(q-i)e^{-\frac{\Gamma}{2}t} + \frac{1}{\pi} \int e^{-i\frac{\Gamma}{2}\epsilon t} \, d\epsilon + \frac{q-i}{\pi} \int \frac{e^{-i\frac{\Gamma}{2}\epsilon t} - e^{-\frac{\Gamma}{2}t}}{\epsilon+i} \, d\epsilon \right] \\
&= c_b^{(0)} \langle \psi_b|z|g \rangle^* e^{-i\Omega_r t} \frac{1}{q^2} \left[q(q-i)e^{-\frac{\Gamma}{2}t} + \frac{4}{\Gamma}\delta(t) + \frac{q-i}{\pi}e^{-\frac{\Gamma}{2}t}2\pi i + \frac{q-i}{\pi}e^{-\frac{\Gamma}{2}t}\pi i \right] \\
&= c_b^{(0)} \langle \psi_b|z|g \rangle^* \frac{1}{q^2} \left[\frac{4}{\Gamma}\delta(t) + (q-i)^2 e^{-i\Omega_r t}e^{-\frac{\Gamma}{2}t} \right].
\end{aligned}
$$

$$(5.36)$$

If we assume that $\langle \psi_b|z|g \rangle$ is real and the expansion coefficient is $c_b^{(0)}$ purely imaginary, then we can finally arrive at the form of the dipole that is reported in [90, 91, 93],

$$
d(t) \propto i \left[2\delta(t) + \frac{\Gamma}{2}(q-i)^2 e^{-i\Omega_r t}e^{\frac{\Gamma}{2}t} \right]. \tag{5.37}
$$

This form of the dipole can be understood intuitively as arising naturally from the two interfering processes that are occurring: direct ionization to the continuum and decay from a discrete state to the continuum with a lifetime of $\hbar/\Gamma$. The first δ-function in equation 5.37 represents direct ionization, and the second term represents decay from the discrete state. A schematic of the dipole after an impulsive XUV pulse is shown in figure 5.3. To demonstrate that this formulation of the dipole in the time domain is compatible with Fano's original derivation, the photoabsorption cross section can be evaluated by

$$
\begin{aligned}
\sigma &= \frac{2\omega}{\epsilon_0 c} \text{Im}\left[\frac{\tilde{d}(\omega)}{\tilde{E}(\omega)} \right] \propto \text{Im}\left[\int_{-\infty}^{\infty} d(t)e^{i\omega t} \, dt \right] \\
&\propto \text{Re}\left[1 + \frac{\Gamma}{2}(q-i)^2 \int_0^{\infty} e^{-\frac{\Gamma}{2}t}e^{i(\omega-\Omega_r)t} \right] \\
&\propto \text{Re}\left[1 + \frac{(q-i)^2}{1-i\epsilon} \right] \\
&\propto \frac{(q+\epsilon)^2}{\epsilon^2+1}.
\end{aligned}
$$

$$(5.38)$$

This is exactly the cross section in equation 5.30 that was derived previously.

The complex coefficient in front of the second term of equation 5.37 arises from the configuration interaction, and its influence on the dipole can be elucidated by expressing it as an exponential,

$$
(q-i)^2 = (q^2+1)e^{i\varphi(q)} \tag{5.39}
$$

where the phase is given by

$$
\varphi(q) = 2\arg(q-i). \tag{5.40}
$$

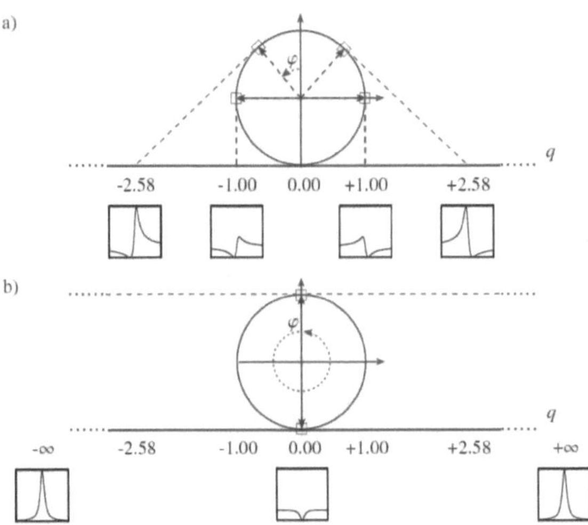

Figure 5.4: Illustration of the mapping between the Fano q parameter and the phase shift φ of the dipole. (a) Phase shift φ is shown as a counter-clockwise rotation of a vector on the unit circle. Dashed tangent lines for a given φ show the mapping to the value $q(\varphi)$ along the q-axis. (b) Shows the special cases of a Lorentzian line shape ($q \to \pm\infty$, $\varphi = 0$) and a window resonance ($q = 0$, $\varphi = \pi$). Adapted from [93].

Substituting this form into the dipole in equation 5.37 gives

$$d(t) \propto i\left[2\delta(t) + \frac{\Gamma}{2}(q^2 + 1)e^{i\varphi(q)}e^{-i\Omega_r t}e^{\frac{\Gamma}{2}t}\right]. \tag{5.41}$$

By expressing the dipole this way, it is clear that the autoionization from the discrete state is phase shifted relative to the instantaneous response (direct ionization). Furthermore, in section 5.2.1 it was established that the q parameter determines the line shape of the photoabsorption cross section ($\sigma \propto \mathrm{Im}[\tilde{d}(\omega)]$). This fact, combined with the correspondence between φ and q, means that the line shape of the Fano resonance is determined by the phase shift of the dipole response. Thus, if one can experimentally introduce a phase shift to the dipole response then it is possible to control the absorption process. In fact, if one has sufficient control over the dipole phase shift then it is possible to continuously change the line shape from symmetric to asymmetric and vice versa. This mapping of q parameter and the phase shift φ and its affect on the line shape is shown in figure 5.4.

5.3 Dipole Control Model

Now that the relationship between the dipole and the absorption line shape has been established in the time domain, we would like to develop a method to control and modify the dipole. This will be done by introducing an IR probe/dressing pulse at a variable time delay τ. This IR pulse will lead to a modification of the dipole that can be modeled analytically for certain assumptions. This model is referred to as the dipole control model (DCM), and it was originally formulated by A. Blätterman, et al. [93–95]. In a similar manner to the previous section, a key assumption that must be made in this model is that, in addition to the XUV pulse, the IR pulse must be a δ-function in time. This is generally only a reasonable approximation if the pulse duration is much shorter than the lifetime $\hbar/\Gamma$. These assumptions will be questionable for the experiments discussed in this chapter, however this model will allow for an understanding of the features that will be seen in the data shown later.

A schematic of the physical situation considered for the DCM is shown in figure 5.5. A XUV pulse at $t = 0$ induces a dipole $d(t)$, and an IR pulse perturbs the dipole after a time delay of τ. The δ-function nature of both pulses means that there are three distinct temporal regions to consider. The first is $t < 0$, and in this region, the dipole response is zero because the XUV pulse hasn't yet populated the excited state. For the region between the two pulses, the dipole is allowed to freely evolve in time, as was derived in 5.41. Thus, the dipole response here is just simply

$$d(0 < t < \tau) = f_0(t) \propto i\left[c\delta(t) + \frac{\Gamma}{2}(q - i)^2 e^{-i\Omega_r t}e^{\frac{\Gamma}{2}t}\right]. \tag{5.42}$$

At $t = \tau$, the dressing field will interact with the system through resonant and non-resonant processes which will modify the dipole response. However, since the dressing field is infinitesimally short in duration, the system will again freely evolve in time for times after τ, but the dressing field is assumed to have modified the dipole amplitude and phase. Therefore, in the third temporal region of $t > \tau$, the dipole response is given by

$$d(t > \tau) = A f_0(t) \tag{5.43}$$

where A is the complex perturbation to the dipole response induced by the dressing pulse. By combing the dipole response for the three temporal regions into a single piece-wise function, we arrive at

$$d_\tau(t) = \begin{cases} 0 & t < 0 \\ f_0(t) & 0 < t < \tau \\ A(\tau)f_0(t) & t > \tau. \end{cases} \tag{5.44}$$

In general, A is a complex quantity that can be explicitly dependent upon τ or ω, and it can be used to describe both resonant and non-resonant processes. For example, a decrease of the amplitude of A can be used to represent ionization of the excited state by the dressing field, whereas the phase can be used to describe a ponderomotive shift of energy levels by the dressing field. To describe these non-resonant processes, it can be assumed that A takes the form of $A = a_1 e^{i\phi_1}$. To describe a resonant process such as coupling of excited states, A takes the form $A = 1 + a_2 e^{i(\Delta\omega\tau + \phi_2)}$ where $\hbar\omega$ is the separation of the states. This form is motivated by the results of perturbation theory [93, 96]. These two primitive forms can be linearly combined into

$$A = a_1 e^{i\phi_1}(1 + a_2 e^{i(\Delta\omega\tau + \phi_2)}). \tag{5.45}$$

All of these quantities can be explicitly dependent upon τ and ω, however they are independent of t given the δ-function pulse durations that are assumed. An illustration of these processes is shown in the level diagram in figure 5.6.

Armed with the dipole in the time domain, it is now possible to calculate the photoabsorption cross section that is the typical observable in an ATS experiment. To do this, one must calculate the spectral dipole in the frequency domain, and this is done by simply

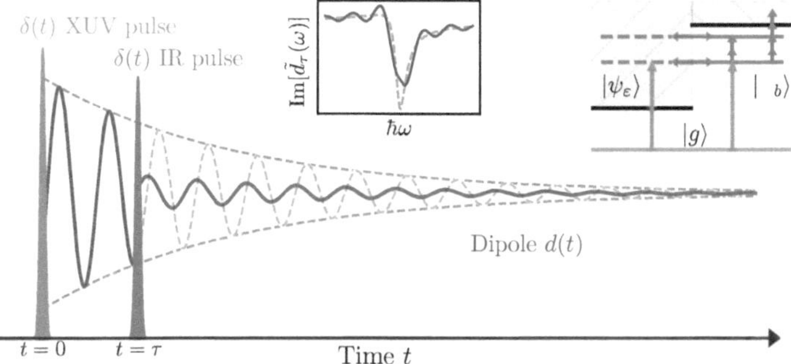

Figure 5.5: Illustration of the dipole response being modified by an impulsive IR pulse. The IR pulse is shown to arrive after a time delay τ. The dipole is allowed to freely evolve between $t = 0$ and $t = \tau$, but after the IR pulse it is modified in amplitude and phase. The effect that this has on the absorption cross section is shown in the central inset for a time delay immediately after τ. Inset on the upper right shows both non-resonant and resonant processes that can be induced with the IR pulse.

Fourier transforming the dipole in equation 5.44,

$$
\begin{aligned}
\tilde{d}_\tau(\omega, \tau) &= \int_{-\infty}^{\infty} d_\tau(t, \tau) e^{i\omega t}\, \mathrm{d}t \\
&= \int_0^\tau f_0(t) e^{i\omega t}\, \mathrm{d}t + \int_\tau^\infty A(\tau) f_0(t) e^{i\omega t}\, \mathrm{d}t \\
&= \int_0^\tau i\left[2\delta(t) + \frac{\Gamma}{2}(q - i)^2 e^{-i\Omega_r t} e^{\frac{\Gamma}{2} t}\right] e^{i\omega t}\, \mathrm{d}t \\
&\quad + A(\tau) \int_\tau^\infty i\left[2\delta(t) + \frac{\Gamma}{2}(q - i)^2 e^{-i\Omega_r t} e^{\frac{\Gamma}{2} t}\right] e^{i\omega t}\, \mathrm{d}t \\
&= i - \gamma(q - i)^2 \frac{1 - (1 - A(\tau)) e^{-\gamma\tau + i\delta\tau}}{\delta + i\gamma}
\end{aligned}
\tag{5.46}
$$

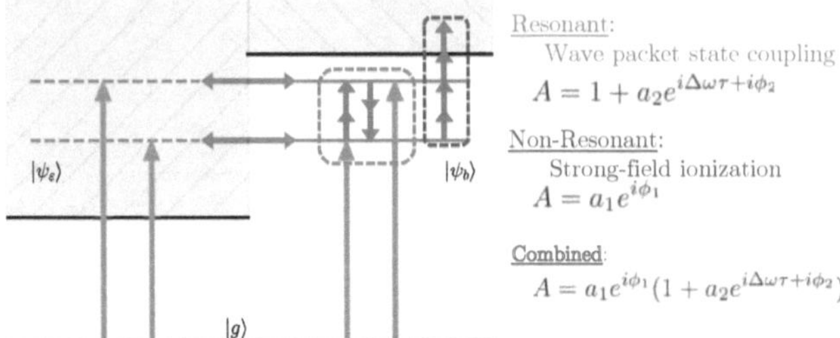

Figure 5.6: Level diagram showing the influence of the IR dressing pulse on a series of Fano resonances. Generally, the IR field can either couple different resonances or directly ionize from the discrete state. Coupling is shown here as a two-photon transition between discrete states, and direct ionization from a discrete state is shown as a multiphoton process. Green arrows represent the configuration interaction coupling the discrete states to the continuum.

where $\gamma = \Gamma/2$ and $\delta = \omega - \omega_r$. The photoabsorption cross section is now given by

$$\begin{aligned}
\sigma(\omega, \tau) &= \frac{2\omega}{\epsilon_0 c} \text{Im}\left[\frac{\tilde{d}(\omega)}{\tilde{E}(\omega)}\right] \propto \text{Im}\left[\int_{-\infty}^{\infty} d(t)e^{i\omega t}\, dt\right] \\
&\propto \text{Im}\left[i - \gamma(q - i)^2 \frac{1 - (1 - A(\tau))e^{-\gamma\tau + i\delta\tau}}{\delta + i\gamma}\right] \\
&= \frac{(q + \delta/\gamma)^2}{(\delta/\gamma) + 1} + \text{Im}\left[\frac{\gamma(q - i)^2(1 - A(\tau))e^{-\gamma\tau + i\delta\tau}}{\delta + i\gamma}\right].
\end{aligned} \tag{5.47}$$

From this, we can now begin to see the effect that of dressing pulse can have on the cross section. For the case of $A(\tau) = 1$, the second term in equation 5.47 goes to zero, and only the unperturbed cross section represented by the first term remains.[10] This is expected because $A(\tau) = 1$ implies that there is no interaction between the dressing field and the system. For the more interesting case of $A(\tau) \neq 1$, the dipole and the cross section will be modulated as a function of both photon energy ω and time delay τ. The specific functional form of $A(\tau)$ will ultimately determine the line shape, and a couple of cases will be discussed herein. It is important to note that as $\tau \to \infty$, the cross section returns to the unperturbed case. This due to the fact that the dipole is allowed to freely evolve over the natural lifetime of the state $\hbar/\Gamma$ before the dressing pulse arrives.

In general, there are two primary cases to consider: resonant and non-resonant processes. For a non-resonant process, it can be assumed that the $A(\tau) = a_1 e^{i\phi_1}$, and the cross section is plotted in figure 5.7(a) assuming the extreme case of $a_1 = \phi_1 = 0$. This

[10]This can be seen by noting that $\epsilon = \delta/\gamma$.

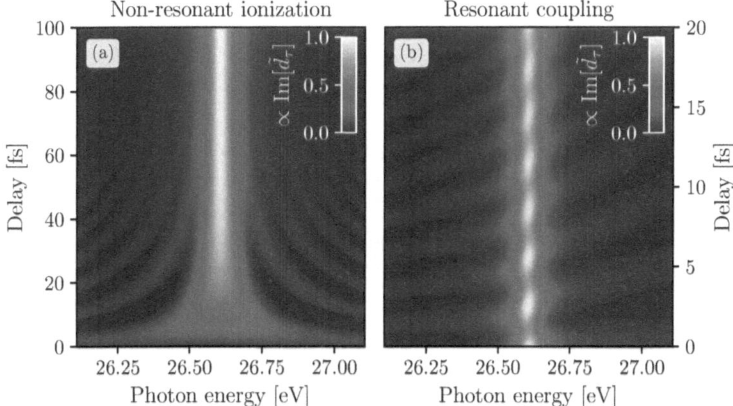

Figure 5.7: Imaginary part of the dipole calculated as a function of photon energy and delay. Resonance position is taken to be 26.605 eV, and the line shape is Lorentzian ($\varphi=0$). (a) Non-resonant case showing complete ionization by dressing field: $a_1 = 1$ and $\phi_1 = 0$. (b) Resonant case showing coupling between states: $a_2 = 0.1$, $\phi_2 = \pi/4$, and $\hbar\Delta\omega = 1.39$ eV.

assumption corresponds to complete ionization of the discrete state by the dressing pulse, and correspondingly, the dipole is quenched for $t > \tau$. Even this extremely simple model is able to reproduce the common hyperbolic features that occur for $\delta\tau = $ const. and have been observed in many transient absorption experiments [77, 81, 92, 97]. These hyperbolic features arise as a physical manifestation of the ringing described by Gibb's phenomenon which occurs because the observable in the measurement (the cross section) is the Fourier transform of a truncated signal (the dipole) [98, 99]. The other dominant feature that can be seen in figure 5.7(a) is a broadening of the resonance width for $t < \hbar/\Gamma$ that is due to the abrupt quenching of the dipole before it is allowed to freely develop over the state's lifetime. The spectral width for this region is proportional to $1/\tau$, and it asymptotically approaches the natural width for $\tau \to \infty$.

For a resonant process, it can be assumed that $A(\tau) = 1 + a_2 e^{\Delta\omega\tau + i\phi_2}$, and the cross section for such a process is plotted in figure 5.7(b) assuming that $a_2 = 0.1$, $\phi_2 = \pi$, and $\hbar\Delta\omega = 1.39$ eV. This form of $A(\tau)$ is motivated by time-dependent perturbation theory, and it can be used to describe coupling of states in a wave packet [92–95]. The idea is that the XUV pulse coherently excites a wave packet of multiple excited states, and then the dressing field couples states through multiphoton transitions. This coupling leads to an oscillation in the cross section's amplitude as a function of delay that is the result of a quantum path interference. This interference can be clearly seen in 5.7(b) with a frequency of $\hbar\Delta\omega = 1.39$ eV at the resonance energy. An additional feature that is present in 5.7(b) is a tilting of the oscillation pattern. This effect arises because the transition frequency is varying across the resonance as a function of photon energy due to the term $e^{i(\delta+\Delta\omega)\tau}$ in equation 5.47 with the resonant form of $A(\tau)$. It is important to emphasize that this description only applies to coupling of states that have been coherently excited because the quantum path interference that gives rise to this oscillation requires a common clock between the coupled states. This is not the case for the dressing field coupling a bright state (one that is initially populated) to a dark state (one that is initially unpopulated). This dark state coupling will only appear as a depletion of the bright state population and is similar to the non-resonant case described previously.

The importance of this simple analytical model is that it allows for an intuitive description of features that will be seen in the ATS experiment that will be described later in this chapter. Additionally, the full complex dipole can be calculated with this model, not just the typical observable that is the cross section ($\propto \mathrm{Im}[\tilde{d}(\omega)]$). This will be used to describe the results in chapter 6 where the full complex dipole will be measured. The complex parts, amplitude, and phase for the non-resonant and resonant cases described above are plotted in figure 5.8.

Further insight into these resonant and non-resonant mechanisms can be gained by Fourier transforming along the time delay axis and examining the characteristic frequencies associated with each process. Specifically, the typical observable (the cross section) is Fourier transformed and this expression is given by

$$\tilde{d}_\nu(\omega,\nu) = \int_{-\infty}^{\infty} \mathrm{Im}[\tilde{d}_\tau(\omega,\tau)]e^{-i\nu\tau}\,\mathrm{d}\tau \propto \int_{-\infty}^{\infty} \sigma(\omega,\tau)e^{-i\nu\tau}\,\mathrm{d}\tau. \tag{5.48}$$

Effectively, this represents two Fourier transforms of the time domain dipole: one that physically occurs by measuring the cross section and another that is performed numerically. This is a two-dimensional spectroscopic representation that will be referred to as the two-dimensional absorption spectrum (2DAS) [94, 95]. The main advantage of analyzing the absorption spectrogram in this manner is that different physical processes are naturally separated because of their differing frequency response.

To see this more clearly, the two general cases of resonant and non-resonant interactions introduced by the IR dressing field will be considered. The first process to consider is the non-resonant case that deals with both direct ionization and energy level shifts by the dressing field, as well as modifications of the dipole phase and q-parameter. These effects are generally described by the complex form $A(\tau) = a_1 e^{i\phi_1}$, and the 2DAS for two sets of a_1 and ϕ_1 parameters is shown in figure 5.9. The immediately striking feature that appears

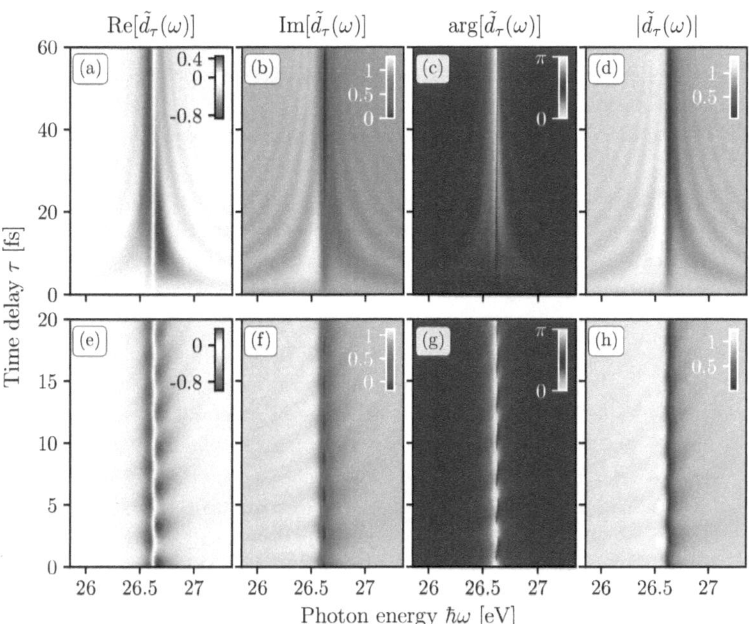

Figure 5.8: Complex parts, phase, and amplitude of $\tilde{d}_\tau(\omega)$ calculated for non-resonant ((a) - (d)) and resonant ((e) - (h)) cases for a resonance at 26.605 eV with $q = -0.258$. Parameters for $A(\tau)$ are $a_1 = 1$ and $\varphi_1 = 0$ for non-resonate and $a_2 = 0.1$, $\phi_2 = \pi/4$, and $\hbar\Delta\omega = 1.39$ eV for resonant.

in $|\tilde{d}_\nu(\omega)|$ are strong lines that original at the resonance photon energy and zero Fourier frequency. There are three lines in total with a vertical line occurring at the resonance energy ω_r and two diagonal lines that for $(\omega - \omega_r) \pm \nu = 0$. The diagonal lines have a slope of one, and they correspond to the hyperbolic features that occur in the cross section because of truncation of the dipole in the time domain by the dressing field. The vertical line that occurs at the resonance energy is related to the bandwidth of the non-resonant process. Since the dressing field is approximated as an instantaneous change of the dipole, the bandwidth in this case extends over all frequencies, however this won't be the case for a finite pulse duration of the dressing field. From the amplitude $|\tilde{d}_\nu(\omega)|$ and phase $\arg[\tilde{d}_\nu(\omega)]$ of these features it is possible to extract both a_1 and ϕ_1 which can characterize the type of non-resonant process that is occurring.

The other case to consider is resonant interactions introduced by the dressing field. This usually comes in the form of coupling bright states populated by the XUV pulse, and, by coupling states within a wave packet in this manner, a fast modulation of the cross section will arise. Within this model, it can be assumed that the form for this type of interaction is given by $A(\tau) = 1 + a_2 e^{i\Delta\omega\tau + i\phi_2}$ where $\Delta\omega$ is the energy separation of the states being coupled and ϕ_2 is the phase that is imparted. This is shown in figure 5.10 for two different coupling strengths and imparted phase shifts. From the Fourier transform of the delay dependent dipole in figures 5.10 (b) and 5.10 (f), it can be seen that these modulations occur at a Fourier frequency corresponding to the energy separation of the states being coupled ($\nu = \Delta\omega$). Also of note are the diagonal structures whose amplitude is center at $(\delta, \nu) = (0, \Delta\omega)$. The diagonal line is given by $\delta + \Delta\omega \pm \nu = 0$, and it effectively "points" to the resonance that is being coupled. This can be seen most clearly in figure 5.11 where the two resonances that are being coupled are shown. Additionally, by looking at the amplitude of the peak at $(\delta, \nu) = (0, \Delta\omega)$ the amplitude of the coupling strength a_2 can be determined because this peak is proportional to a_2. The effect of the imparted phase ϕ_2 can be seen clearly by looking at the phase of the Fourier transform ($\arg[\tilde{d}_\nu(\omega)]$) in figures 5.10 (c) and 5.10 (g). The overall structure between the two cases is the same, but the actual phase shifts across these structures is modified by the phase that is imparted by the dressing field. From the amplitude $|\tilde{d}_\nu(\omega)|$ and phase $\arg[\tilde{d}_\nu(\omega)]$ of these features it is possible to extract both a_2, $\Delta\omega$, and ϕ_2 which characterizes the resonant interaction.

5.3.1 Light-Induced Phase

In the previous section, the parameters that characterized the modification to the dipole in the DCM were kept as general as possible and just treated as parameters that could be delay dependent. In this section and the next, specific forms for the non-resonant parameters ϕ_1 and a_1 will be introduced, and this treatment will allow for an extension of the DCM to dressing fields with a finite pulse duration. This section will discuss the light-induced phase

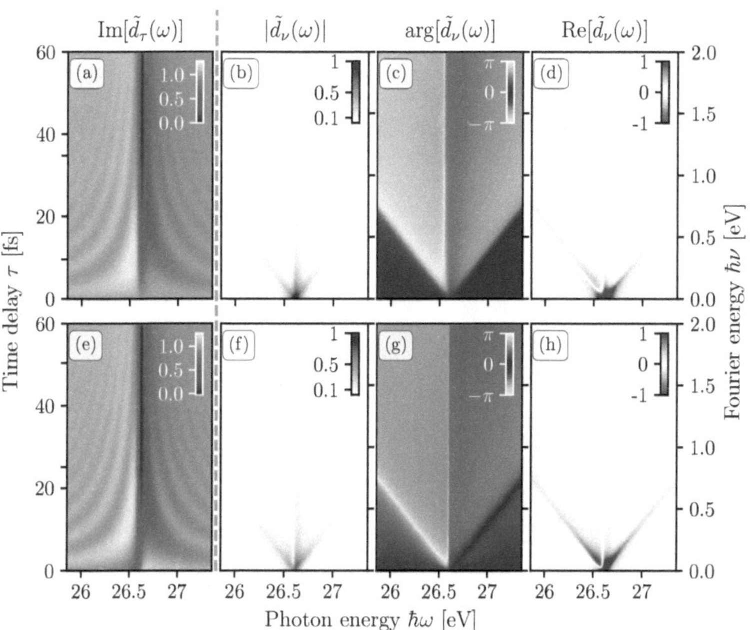

Figure 5.9: Fourier transform of the dipole, $\tilde{d}_\nu(\omega)$, for the non-resonant interaction case where $A(\tau) = a_1 e^{i\phi_1}$. Resonance shown here has a resonance energy of 26.605 eV and a q-parameter of -0.258. Figures (a)-(d) are for complete ionization by the dressing field, and this corresponds to $a_1 = 0, \phi_1 = 0$. Figures (e)-(h) are for partial ionization with a phase shift, and this corresponds to $a_1 = 0.8, \phi_1 = \pi/2$.

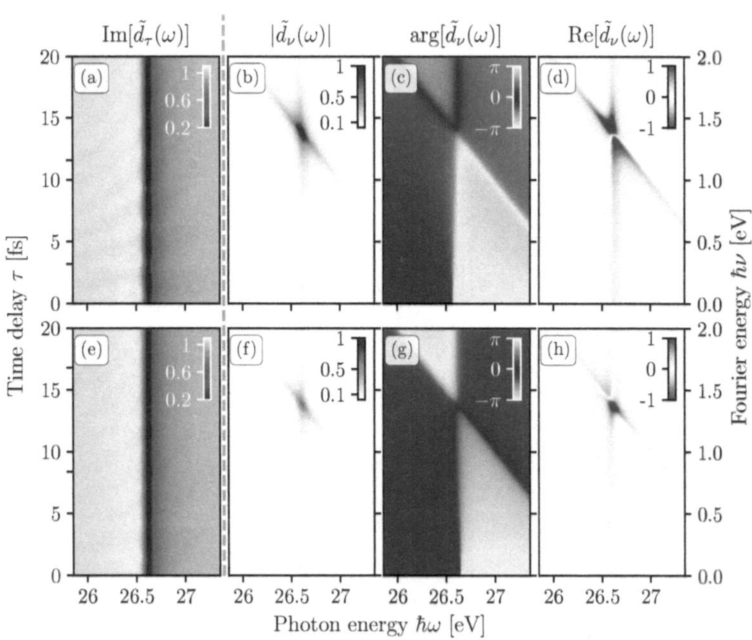

Figure 5.10: Fourier transform of the dipole, $\tilde{d}_\nu(\omega)$, for the resonant interaction case where $A(\tau) = 1 + a_2 e^{i\Delta\omega\tau + i\phi_2}$. Resonance shown here has a resonance energy of 26.605 eV and a q-parameter of -0.258. Figures (a)-(d) are for a coupling strength and phase shift give by $a_2 = 0.1, \phi_1 = \pi$. Figures (e)-(h) are for a coupling strength and phase shift give by $a_2 = 0.05, \phi_1 = \pi/2$.

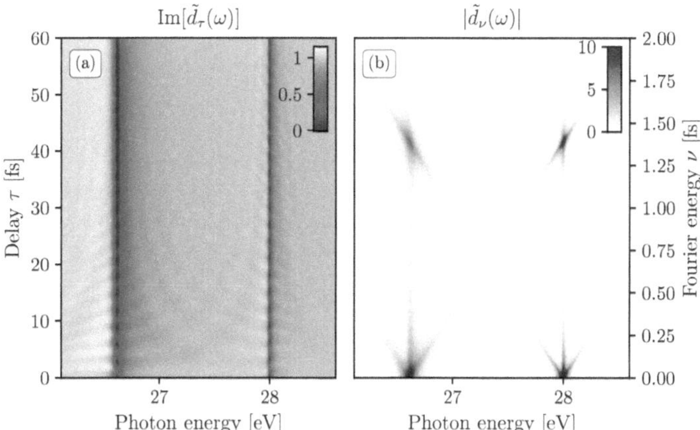

Figure 5.11: Fourier transform of the dipole, $\tilde{d}_\nu(\omega)$, for resonant and non-resonant interactions with two resonances. $\phi_0 = 4p \; a_2 = 0.1, \phi_2 = \pi, \, a_2 = 0.05, \phi_2 = \pi/2$

(LIP) whose effect is captured by ϕ_1.

In section 5.2.2, the connection between the dipole phase and the absorption lineshape was established through the relationship given in equation 5.40. In the DCM, the natural phase of the resonance is modified by the phase introduced through the parameter ϕ_1. Thus, the LIP given by ϕ_1 controls the absorption lineshape of the bright resonance. In very general terms, this LIP originates from the energetic Stark shift $\Delta\varepsilon(t,\tau)$ of a bright state that is the result of coupling to nearby states. This LIP can be generally calculated using the relationship

$$\phi_1(t,\tau) = \frac{1}{\hbar} \int_0^t \Delta\varepsilon(t',\tau)\,\mathrm{d}t' \tag{5.49}$$

where the assumption is made that the XUV pulse is still a delta function in time and populates the bright state at a time $t = 0$.[11] This description of the total accumulated phase allows for dressing pulses that have finite pulse duration. Thus, the challenge now is to calculate the Stark shift $\Delta\varepsilon(t,\tau)$ for a given system.

There are several models to calculate this energy shift, and three of these will be discussed herein. The first method is to consider the electron in the bright state as a quasi-free electron that couples very weakly to nearby states and is weakly bound. This assumption means that the energy shift will be related to the ponderomotive energy that the electron

---

[11] Otherwise, the integration limits would extend from $-\infty$ to t.

picks up from the field. The energy shift can be calculated classically as

$$\Delta\varepsilon(t,\tau) = \frac{1}{2}mv(t,\tau)^2 = \frac{e^2}{2m}\left[\int_{-\infty}^{t} \mathcal{E}(t',\tau)\,dt'\right]^2 \tag{5.50}$$

where $\mathcal{E}(t,\tau)$ is the electric field of the dressing pulse. This method was used to describe several experiments that were performed in He [91, 100].

The second method to calculate the energy shift is to use second order time dependent perturbation theory [101–104]. This method is a sub-cycle extension of the optical AC Stark shift, where the energy shift $\Delta\varepsilon_a$ of state a is given by

$$\Delta\varepsilon_a = -\frac{1}{2}\sum_{k\neq a}\frac{\omega_{ka}|d_{ka}|^2}{\omega_{ka}^2 - \omega_L^2}\langle E(t)^2\rangle = -\frac{1}{2}\alpha_a\langle E(t)^2\rangle \tag{5.51}$$

where α_a is the polarizability of state a, d_{ka} is the dipole matrix element between states a and k, ω_L is the dressing frequency, and $E(t) = E_0(t)\cos(\omega_L t)$ is the electric field. This is an inherently cycle averaged effect, but it can be extended to a sub-cycle ac Stark shift using second order perturbation theory. The energy shift for this case is given by

$$\Delta\varepsilon_a(t,\tau) = -i\sum_{k\neq a}d_{ak}\mathcal{E}(t-\tau)e^{i\omega_{ak}t}\int_{-\infty}^{t} d_{ka}\mathcal{E}(t'-\tau)e^{i\omega_{ka}t}\,dt'. \tag{5.52}$$

For a pulse shape that is given by $\mathcal{E}_0(t) = \mathcal{E}_p e^{-|t|/t_p}$, the integrals can be solved analytically to give

$$\Delta\varepsilon_a(t,\tau) = \frac{1}{2}\mathcal{E}_0(t-\tau)^2\sum_{k\neq a}\left[\frac{\omega_{ka}|d_{ka}|^2}{\omega_{ka}^2 - \omega_L^2}\cos(\omega_L(t-\tau))^2 - i\frac{\omega_L|d_{ka}|^2}{\omega_{ka}^2 - \omega_L^2}\sin(2\omega_L(t-\tau))\right]$$

$$= \frac{1}{2}\mathcal{E}_0(t-\tau)^2[\alpha_a\cos(\omega_L(t-\tau))^2 - i\gamma_a\sin(2\omega_L(t-\tau))]. \tag{5.53}$$

where α_a is the polarizability of state a and γ_a is related to the population change in state a as it is being coupled to state k. This form for the energy level shift is generally only accurate when the frequency of the dressing field is far from resonance and the peak Rabi frequency $\Omega_{ka} = d_{ka}E_p$ is much less than the energy level separation.

The third method to calculate the energy level shift from the dressing field involves the use of Floquet theory and the rotating-wave approximation (RWA)[97]. Floquet theory will be discussed in more detail in section 5.4 to describe light-induced states, but it will be briefly discussed here as a means to derive the energy shift of a bright state in a simple two-level system consisting of a bright state $|1\rangle$ and a dark state $|2\rangle$ with energies of ω_1 and ω_2 in atomic units. In the Floquet basis, the wave function can be expanded in such a way that the time evolution of the dressed states is simply a phase gain,

$$|\psi(t)\rangle = \sum_{\alpha,n} C_\alpha e^{-i(\varepsilon_\alpha + n\omega_L)t} |\phi_\alpha, n\rangle. \tag{5.54}$$

where $|\phi_\alpha, n\rangle$ and $C_\alpha = \langle \phi_{\alpha,n}|1\rangle$ are the Floquet states and initial amplitude, respectively. The amplitude in the bright state is then given by

$$\begin{aligned}
C_1(t) = \langle 1|\psi(t)\rangle &= \sum_{\alpha,n} \langle 1|\phi_{\alpha,n}\rangle \langle \phi_{\alpha,n}|1\rangle \, e^{-i(\varepsilon_\alpha + n\omega_L)t} \\
&\approx \langle 1|\phi_{+,0}\rangle \langle \phi_{+,0}|1\rangle \, e^{-i\varepsilon_+ t},
\end{aligned} \tag{5.55}$$

where in the last step the approximation was made that $\langle \phi_{\alpha,n}|1\rangle \approx 0$ except for $\langle \phi_{+,0}|1\rangle$. It can be seen from this relation that the Floquet theory naturally agrees with the LIP picture where the bright state is modified only in an accumulation of phase. Therefore, the energy shift that can be used to calculate the LIP is given by

$$\Delta\varepsilon = \varepsilon_+ - \omega_1 \tag{5.56}$$

where ω_1 is the energy of the bright state. Thus, the problem has been reduced to calculating the Floquet energy ε_+. This calculation can be done using the RWA to calculate the first-order Floquet energy, and this is shown in the review by Wu, et al. [97]. The resulting expression for the Floquet energy is given by

$$\Delta\varepsilon_+ = \omega_1 + \frac{-\Delta + \sqrt{\Delta^2 + \Omega(t,\tau)^2}}{2}, \tag{5.57}$$

where $\Delta = \omega_L - (\omega_2 - \omega_1)$ is the detuning of the dressing frequency from resonance and $\Omega(t,\tau) = d_{21}\mathcal{E}(t,\tau)$ is the Rabi frequency. Thus, the time-dependent energy level shift is given by

$$\Delta\varepsilon(t,\tau) = \frac{-\Delta + \sqrt{\Delta^2 + \Omega(t,\tau)^2}}{2}. \tag{5.58}$$

The advantage of this method to calculate the energy level shift is that it does not require the dressing frequency to be far from resonant, and this method works best when the detuning is small. This is due to the use of the RWA approximation in this derivation, and it can be generally shown that the RWA is most appropriate when the detuning is small and the dynamics are dominated by Rabi cycling [97, 105].

5.3.2 Light-Induced Attenuation

An additional term in the DCM that can be explicitly calculated for a dressing pulse of finite pulse duration is the non-resonant amplitude a_1, which, as stated previously, is related to the population of the discrete state. For typical experimental conditions that will be considered

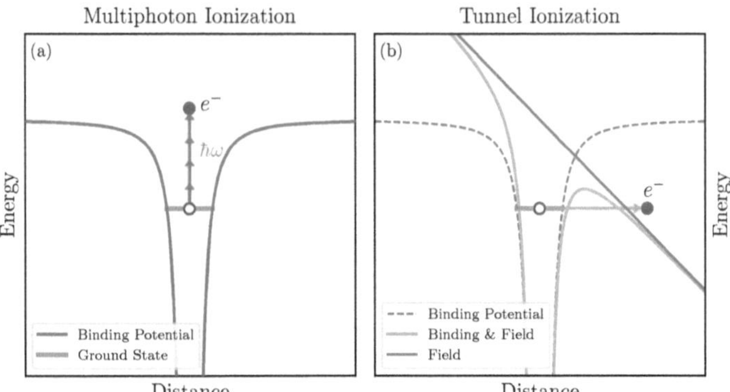

Figure 5.12: Schematic of multiphoton and tunnel ionization of an atom in a strong laser field. (a) Multiphoton ionization regime where the electron can absorb multiple photons to reach the continuum and be ionized. (b) Tunnel ionization regime where the electric field of the dressing laser distorts the atomic binding potential to allow for bound electron to tunnel through the finite barrier created by the combined atom-field potential.

herein, the dressing pulse has a large enough field strength that is can drive ionization of the discrete state to the continuum even when the photon energy of the dressing pulse is insufficient to directly ionize. This means that the amplitude a_1 will be proportional to the ionization probability of the discrete state. In this section, the two main regimes of ionization by a strong field will be discussed, and the amplitude a_1 will be calculated. The two regimes to consider are multiphoton and tunneling ionization, and they can be thought of as opposite extrema of photoionization by a strong field.

In the multiphoton regime, the ionization process can be treated perturbatively for low intensity, and the physical picture is that the electron is able to absorb multiple photons to gain enough energy to reach the continuum, as shown in figure 5.12 (a). The perturbative treatment means that the ionization probability will be proportional to I^N where I is the intensity of the dressing field and N is the minimum number of photons required to reach the continuum [11]. Absorption of additional photons beyond the minimum number required to ionize is also possible and is know as above threshold ionization (ATI), and was first observed experimentally in 1979 [106].

In the limit of longer wavelengths, the nonlinear order N increases, and eventually, another process begins to dominate. Namely, the ionization process in this limit begins to resemble ionization by a static electric field. This is schematically shown in figure 5.12 (b), and this physically corresponds to suppression of the Coulomb binding potential to the point that an electron has a reasonable probability of tunneling through the finite barrier that is created by the combined atomic and electric fields. The low frequency (long wavelength) allows this to happen by giving the electron enough time to tunnel through the barrier each half-cycle of the dressing laser period. In this regime, the ionization probability no longer follows the perturbative scaling of I^N, and it instead follows an exponential scaling in field given by $\exp(-2F_0/3\mathcal{E})$ where F_0 is related the binding electric field and $\mathcal{E}$ is the dressing electric field [89, 107].

To delineate between these two extrema regimes of ionization, a useful parameter is the Keldysh parameter

$$\gamma = \sqrt{\frac{I_p}{2U_p}} \propto \sqrt{\frac{I_p}{I_0\lambda^2}} \tag{5.59}$$

where I_p is the ionization potential, U_p is the ponderomotive energy, I_0 and λ are the laser intensity and wavelength [107]. The Keldysh parameter represents the adiabiticity of the process, and it determines which regime should be considered. For strong fields at long wavelengths ($\gamma \ll 1$), the electron is able to tunnel ionize each half-cycle of the field because the rate of change of the field is slow and the tunneling rate is high. For weaker fields at shorter wavelengths ($\gamma \gg 1$), the electron has to absorb multiple photons to gain the necessary energy to ionize.

There are generally two analytic methods to calculate the ionization rate in a strong field, PPT [108] and ADK [109].[12] The full details of their derivation will not be reproduced here, but the main results that are pertinent will be discussed. The PPT model is the more general of the two, and it is generally applicable to calculating the ionization rate and probability for both regimes of strong field ionization. It assumes a hydrogenic atom and uses effective quantum numbers to generalize to other atoms. Following the formalism found in [89], the rate of ionization w in the PPT model is

$$w_{PPT}(\mathcal{E},\omega) = \sum_{q \geq q_{th}}^{\infty} w_q(\mathcal{E},\omega) \tag{5.60}$$

where w_q is the rate for absorbing q photons and $q_{th} = \lceil (I_p+U_p)/\omega \rceil$ is the minimum number of photons required to ionize an electron after including the AC Stark effect. Assuming the electron is in a state with quantum numbers n, l, and m, the total rate can be written as

$$\begin{aligned}
w_{PPT}(\mathcal{E},\omega) = &|C_{n^*l^*}|^2 G_{lm} I_p \left(\frac{2F_0}{\mathcal{E}}\right)^{2n^*-|m|-1} \left(\frac{1}{\sqrt{1+\gamma^2}}\right)^{-|m|-1} \\
&\times \frac{4}{|m|\sqrt{3\pi}} \left(\frac{\gamma^2}{1+\gamma^2}\right) e^{-\frac{2F_0}{3\mathcal{E}}g(\gamma)} \sum_{q \geq q_{th}}^{\infty} A_q(\omega,\gamma)
\end{aligned} \tag{5.61}$$

<hr>

[12]The acronyms stem from the last names of the authors.

105

where $\mathcal{E}$ is the electric field of the laser, F_0 is the binding field strength related to I_p, γ is the Keldysh parameter, $|C_{n^*l^*}|^2 G_{lm}$ is a constant related to the atom, and the functions $g(\gamma)$ and $A_q(\omega, \gamma)$ are given in [89]. Generally speaking, $g(\gamma)$ is a function that approaches unity as $\gamma \to 0$, and $A_q(\omega, \gamma)$ falls off exponentially with q and increases exponentially with $\mathcal{E}$ up to a saturation point related to I_p. In principle this method has a significant advantage in the fact that it is applicable for both regimes of ionization, however evaluation of the infinite sum makes this method computationally expensive. The ionization rate for several atomic species is shown in figure 5.13 (a).

The other main method that is commonly used to calculate the ionization rate is the ADK method [109]. This method can be thought of as an extension of the PPT model in the limit of a static field ($\omega \to 0$), where tunneling ionization is the dominant mechanism. In this limit, all γ-dependent terms become unity and the resulting rate is given by

$$w_{ADK} = |C_{n^*l^*}|^2 G_{lm} I_p \left(\frac{2F_0}{\mathcal{E}} \right)^{2n^* - |m| - 1} e^{-\frac{2F_0}{\mathcal{E}}}. \tag{5.62}$$

This much simpler expression is generally only valid for values γ smaller than roughly 0.5, however the ADK model is used in the literature well beyond this range even though it is known to underestimate the ionization rate [110]. This is mostly due to the ease of computation when compared to PPT. The ADK ionization rate is given for several atomic species in figure 5.13 (a).

Generally speaking, both methods have very similar dependence on field strength, and their exponential behaviour means that the ionization rate is limited to a small time window around the peak of the electric field every half-cycle of the laser pulse. The total ionization probability at a given time t is given by the relation

$$P(t) = 1 - \exp\left(-\int_{-\infty}^{t} w(t') \, dt' \right), \tag{5.63}$$

and examples of the ionization probability integrated over an entire pulse are shown in figure 5.13 (b) for several atomic species.

Returning to the DCM parameter a_1, it is related to the ionization probability $P(t)$ because the strong field is assumed to deplete the population of the excited discrete state. The choice of model is dependent upon the dressing laser parameters and the states being considered, but for either model, a_1 is given by

$$a_1(t, \tau) = \exp\left(-\int_0^t w(t' - \tau) \, dt' \right). \tag{5.64}$$

This is assuming that the XUV pulse is a δ-function centered at $t = 0$. Effectively, this amplitude attenuates the dipole as a function of time based upon the ionization probability by the dressing pulse, hence this effect is referred to as light-induced attenuation (LIA) of

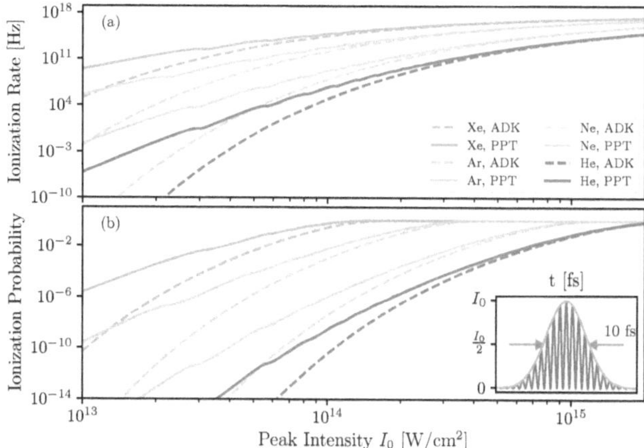

Figure 5.13: (a) Ionization rate calculated for various noble gases using both PPT and ADK models as a function of peak intensity for an 800 nm pulse. (b) Ionization probability integrated through the 800 nm pulse shown in the inset figure as a function of its peak intensity.

the dipole [101, 102, 111].

Since the dressing pulse is no longer treated as a δ-function, the DCM must be modified to include the form of a_1 given in equation 5.64. To do this, the initial assumption for the dipole in equation 5.44 is extended to include the decrease in amplitude throughout the duration of the dressing pulse, and this is given by

$$d(t,\tau) \propto \begin{cases} 0 & t < 0 \\ i\left[\delta(t) + a_1(t,\tau)\frac{\Gamma}{2}(q-i)^2 e^{-i\Omega_r t}e^{-\frac{\Gamma}{2}t}\right] & t > 0. \end{cases} \tag{5.65}$$

The cross section can then be calculated by Fourier transforming the dipole and taking the imaginary part. The dipole in the frequency domain is given by

$$\tilde{d}(\omega,\tau) = \int_{-\infty}^{\infty} d(t,\tau)e^{i\omega t}\, dt \propto i + i\frac{\Gamma}{2}(q-i)^2 \int_0^{\infty} a_1(t,\tau)e^{-i(\omega-\Omega_r)t}e^{-\frac{\Gamma}{2}t}\, dt, \tag{5.66}$$

and the cross section is then

$$\sigma(\omega,\tau) \propto \text{Im}[\tilde{d}(\omega,\tau)] \propto 1 + \text{Im}\left[i\frac{\Gamma}{2}(q-i)^2 \int_0^{\infty} a_1(t,\tau)e^{-i(\omega-\Omega_r)t}e^{-\frac{\Gamma}{2}t}\, dt\right]. \tag{5.67}$$

107

The LIA cross section derived here will be used to explain several features in the ATS spectrogram in section 5.5.3.

5.4 Floquet Theory: Light-Induced States

A feature that is seen in ATS experiments that is not included in the DCM introduced in the previous section 5.3 are light-induced states (LIS) that arise during temporal overlap between the dressing and XUV pulses. These were first seen in early ATS experiments in helium [112, 113]. In [112], the LIS were explained in terms of a Raman-like two-photon process involving the absorption of an XUV photon and either the emission or absorption of a dressing photon to reach a dark state that is dipole-forbidden from the ground state by a single XUV photon. In this case, the LIS would appear in the absorption spectrogram at an energy corresponding to ± 1 dressing photon energy from the dark state.

The description of LIS in terms of a Raman-like process starts to break down when the dressing photon energy is close to resonant to a transition between a bright and dark state [97]. In this regime, it is more appropriate to think of the LIS as arising from the dressed states of the atom in a strong IR dressing field. To describe this effect, Floquet theory will be introduced in this section to describe a strongly driven two-level system consisting of near resonantly coupled bright and dark states [97, 114, 115].

To begin, a two-level system consisting of a bright state $|1\rangle$ and a dark state $|2\rangle$ energetically separated by ω_0 is driven by a periodic electric field at fixed delay. The Hamiltonian of this system is given by

$$\hat{H}(t) \rightarrow \begin{bmatrix} -\frac{\omega_0}{2} & \Omega(t) \\ \Omega(t) & \frac{\omega_0}{2} \end{bmatrix} \tag{5.68}$$

where $\Omega(t) = \mu \mathcal{E}_0 \cos(\omega t) = \Omega_0 \cos(\omega t)$ is the time-dependent Rabi frequency. Since this Hamiltonian is periodic, $H(t + 2\pi/\omega) = H(t)$, then Floquet theory states that the wave function can be expanded in a periodic basis to take the form

$$|\psi(t)\rangle = \sum_f c_f e^{-i\epsilon_f t} |\phi_f(t)\rangle \tag{5.69}$$

where c_f is a time-independent expansion coefficient and $|\phi_f(t)\rangle$ and ϵ_f are the Floquet states and energies [114]. The Floquet states are themselves periodic and satisfy $|\phi_f(t + 2\pi/\omega)\rangle = |\phi_f(t)\rangle$. The importance of representing the wave function in this basis is the fact that the dynamics have been reduced to a trivial phase evolution across each Floquet state that comprises the total wave function. Thus, solving the time-dependent Schrödinger equation in this basis simply consists of finding the Floquest states and their energies. This is done by solving the Floquet equations, which are

$$\left(\hat{H}(t) - i\frac{\partial}{\partial t}\right)|\phi_f(t)\rangle = \hat{H}_F|\phi_f(t)\rangle = e^{-t\epsilon_f t}|\phi_f(t)\rangle \tag{5.70}$$

where $\hat{H}_F$ is the Floquet matrix.

To diagonalize the Floquet matrix, a product basis $|\alpha, n\rangle = |\alpha\rangle \otimes |n\rangle$ is used that consists of the product of the bare state basis $|\alpha\rangle = \{|1\rangle, |2\rangle\}$ and the Fourier basis $|n\rangle$. The Fourier basis states are given by

$$|n\rangle = e^{-in\omega t}$$
$$\langle n|f(t)\rangle = \frac{1}{T}\int_0^T e^{in\omega t} f(t)\, \mathrm{d}t, \tag{5.71}$$

and they represent the Fourier transform of a function $f(t)$ over one period $T = 2\pi/\omega$. Furthermore, since ω is the driving frequency of the dressing field, the values of n take on integer values and represent photon number. In this basis, the Floquet equation 5.70 can be represented by

$$\sum_{\beta,m}\langle\alpha, n|\hat{H}_F|\beta, m\rangle\langle\beta, m|\phi_{\gamma,l}\rangle = q_{\gamma,l}\langle\alpha, n|\phi_{\gamma,l}\rangle. \tag{5.72}$$

The matrix elements $\langle\alpha, n|\hat{H}_F|\beta, m\rangle$ are related to the Fourier transform of the Hamiltonian, $\hat{H}^{[n]} = \langle n|\hat{H}(t)\rangle$, and they are given by

$$\langle\alpha, n|\hat{H}_F|\beta, m\rangle = \hat{H}^{[n-m]}_{\alpha,\beta} + n\omega\delta_{\alpha,\beta}\delta_{n,m}. \tag{5.73}$$

For the two-level system under consideration, the matrix representation of the Floquet Hamiltonian $\hat{H}_F$ in this basis is block tri-diagonal, and it is given by

$$\hat{H}_F \rightarrow \begin{bmatrix} -2\omega-\frac{\omega_0}{2} & 0 & 0 & \frac{\Omega_0}{2} & 0 & 0 & 0 & 0 & 0 & 0 \\ 0 & -2\omega+\frac{\omega_0}{2} & \frac{\Omega_0}{2} & 0 & 0 & 0 & 0 & 0 & 0 & 0 \\ 0 & \frac{\Omega_0}{2} & -\omega-\frac{\omega_0}{2} & 0 & 0 & \frac{\Omega_0}{2} & 0 & 0 & 0 & 0 \\ \frac{\Omega_0}{2} & 0 & 0 & -\omega+\frac{\omega_0}{2} & \frac{\Omega_0}{2} & 0 & 0 & 0 & 0 & 0 \\ 0 & 0 & 0 & \frac{\Omega_0}{2} & -\frac{\omega_0}{2} & 0 & 0 & \frac{\Omega_0}{2} & 0 & 0 \\ 0 & 0 & \frac{\Omega_0}{2} & 0 & 0 & \frac{\omega_0}{2} & \frac{\Omega_0}{2} & 0 & 0 & 0 \\ 0 & 0 & 0 & 0 & 0 & \frac{\Omega_0}{2} & \omega-\frac{\omega_0}{2} & 0 & 0 & \frac{\Omega_0}{2} \\ 0 & 0 & 0 & 0 & \frac{\Omega_0}{2} & 0 & 0 & \omega+\frac{\omega_0}{2} & \frac{\Omega_0}{2} & 0 \\ 0 & 0 & 0 & 0 & 0 & 0 & 0 & \frac{\Omega_0}{2} & 2\omega-\frac{\omega_0}{2} & 0 \\ 0 & 0 & 0 & 0 & 0 & 0 & \frac{\Omega_0}{2} & 0 & 0 & 2\omega+\frac{\omega_0}{2} \end{bmatrix}. \tag{5.74}$$

Diagonalizing this matrix enables one to calculate the Floquet energies and states, and this enables the wave function to calculated for any time. This diagonalization has been done for a system with an energy separation of $\hbar\omega_0 = 0.875$ eV and photon energy $\hbar\omega = 0.867eV$, and the Floquet energies as a function of Rabi frequency Ω_0 is shown in figure 5.14. As can be seen, the Floquet energies are given by two 'ladders' of states separated by $\hbar\omega$ that are

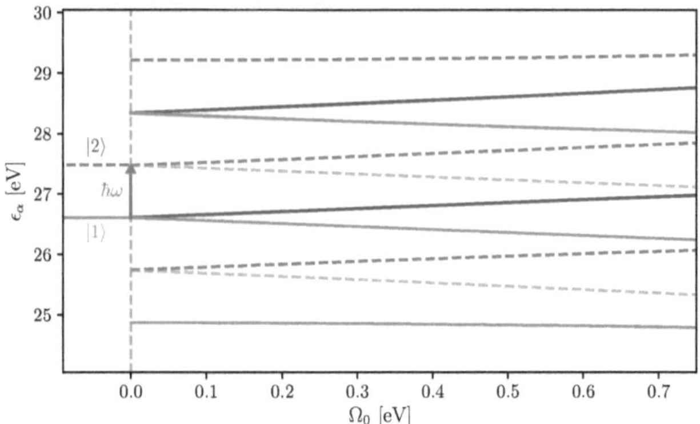

Figure 5.14: Floquet energies as a function of Rabi frequency. System consists of a bright state $|1\rangle$ at a energy of 26.605 eV and a dark state $|2\rangle$ at 27.48 eV, and the dressing field of wavelength 1430 nm (shown as a red arrow). The states which are dipole forbidden from the ground state are shown with dashed lines. Floquet energies are given by two 'ladders' of states separated by $\hbar\omega$ that are centered on each state.

centered on each state.

To connect this to ATS experiments, delay-dependence of the Floquet states must be introduced to calculate the delay-dependent dipole moment. For the constant field envelope that we have been considering thus far, the delay-dependent Floquet states are given by

$$|\psi_\alpha(t,\tau)\rangle = e^{-i\epsilon_\alpha t} \sum_n e^{-in\omega(t+\tau)} |\phi_{\alpha,n}\rangle \tag{5.75}$$

where the Floquet states $|\phi_{\alpha,n}\rangle$ and energies ϵ_α are calculated by diagonalizing $\hat{H}_F$ at a fixed phase $\omega\tau = 0$. Thus, the time- and delay-dependent Floquet states $|\psi_\alpha(t,\tau)\rangle$ can be thought of as a superposition of dressed states that were excited at $t = 0$ with a fixed phase of $\omega\tau$. For an XUV pulse that can be approximated as a δ-function in time, the initial wave function at $t = 0$ is given by the superposition

$$|\Psi(0,\tau)\rangle = \sum_\alpha C_\alpha(\tau) |\psi_\alpha(0,\tau)\rangle, \tag{5.76}$$

where $C_\alpha(\tau) = \langle\psi_\alpha(0,\tau)|\hat{\mu}_X|g\rangle = \sum_n e^{in\omega\tau} \langle\phi_{\alpha,n}|\hat{\mu}_X|g\rangle$ is the dipole transition matrix element from the ground state $|g\rangle$. The wave function for $t > 0$ is now given by

$$|\Psi(t,\tau)\rangle = \sum_\alpha C_\alpha(\tau) |\psi_\alpha(t,\tau)\rangle \,, \tag{5.77}$$

and the time-dependent dipole is

$$d(t,\tau) = \sum_{\alpha,m,n} e^{-i(\epsilon_\alpha + m\omega)t} e^{i(n-m)\omega\tau} \langle \phi_{\alpha,n}|\hat{\mu}_X|g\rangle \langle g|\hat{\mu}_X|\phi_{\alpha,m}\rangle + c.c. \tag{5.78}$$

From this dipole moment many features that are seen in common ATS spectrograms can be easily understood. First of all, the exponential term $e^{i(\epsilon_\alpha + n\omega)t}$ means that the dipole will oscillate at frequencies that corresponds to the dressed states of the atom, and therefore, absorption will be observed at photon energies that correspond to the dressed states which are dipole allowed from the ground state. Thus, the LIS that are seen in ATS experiments can be naturally interpreted as probing of the dressed atom. A second common feature which can be understood from the dipole in equation 5.78 are delay dependent oscillations which arise from the exponential term $e^{i(n-m)\omega\tau}$. Since only dipole allowed dressed states are probed, this means that only every other state in each Floquet ladder will contribute to the dipole. Thus, $n-m$ is even, and the delay-dependent oscillations will be at frequencies of 2ω, 4ω, etc. The 2ω oscillations can be interpreted as quantum path interferences between coherently excited dressed states that are separated by 2ω and end up in the same final state [116]. A schematic of the apperance of LIS in the spectrogram is shown in figure 5.15 where the Floquet matrix for the two-level system considered previously is diagonalized at each time point and the dressed states are shown.

5.5 Strong-field Transient Absorption in Argon

Now that the theoretical background has been established for studying the dynamics of autoionizing states in a strong dressing field, we can turn our attention to the experimental study that was conducted using ATS. The intent of this work was to establish a strong understanding of the dynamics that we should expect to see in chapter 6, where a novel experimental scheme is used to extract a complete picture of the laser-induced dynamics by experimentally measuring the full complex refractive index.

5.5.1 Experimental setup

The autoionizing states that will be studied in this work are the $3s3p^6np$ states in argon. The relevant level diagram is shown in figure 5.1, and a table of the resonance parameters is shown in table 5.1. This series of states were chosen because they are located energetically in a regime that is easily accessible by our HHG sources. The ground state photoabsorption cross section for these states can be calculated using Fano's original theory, equation 5.30,

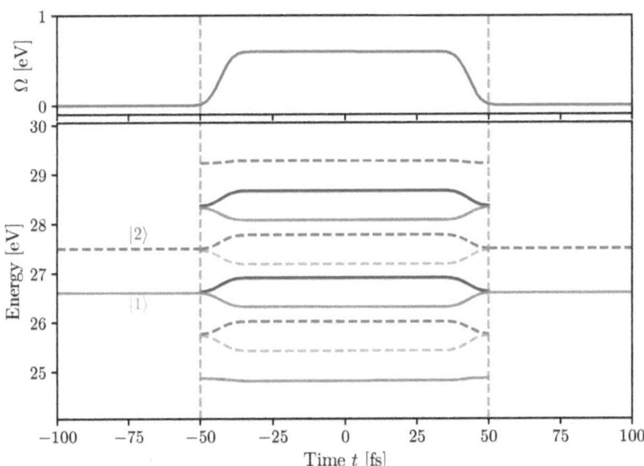

Figure 5.15: (a) Rabi frequency as a function of time that represents the strength of the dressing pulse used in this calculation. (b) Floquet energies as a function of time for the pulse shown in (a). The dressed states which are dipole allowed from the ground state are shown with solid lines, and the dipole forbidden states are shown with dashed lines. The color corresponds to the two initial states: $|1\rangle$ (bright) and $|2\rangle$ (dark).

	E_r [eV]	Γ [eV]	q	ρ^2
$3s3p^64p$	26.605	80.2(7)	-0.286(4)	0.840(3)
$3s3p^65p$	27.994	28.5(8)	-0.177(3)	0.848(3)
$3s3p^66p$	28.509	12.2(3)	-0.135(9)	0.852(9)
$3s3p^67p$	28.757	6.6(1)	-0.125(4)	0.846(9)
$3s3p^68p$	28.898	4.5(2)	-0.132(4)	0.77(2)

Table 5.1: Parameters of the $3s3p^6np$ Fano resonances in argon. These values were extracted from experimental cross sections, see [117–119].

and this is plotted for the resonance parameters of interest in figure 5.16. The non-resonant background absorption used in the calculation for figure 5.16 is interpolated from scattering factors available from CXRO [71]. The imaginary part of the scattering factor is related to the photoabsorption cross section by the relationship

$$\sigma_{\mathrm{NR}} = 2r_0\lambda f_2, \tag{5.79}$$

where r_0 is the classical electron radius and f_2 is the imaginary part of the scattering factor. Since the atomic scattering factor from CXRO does not include fine resonance structure, it can only be used as an approximation of the non-resonant contribution to the total cross section.

The TABLe is the experimental apparatus that will be used in all of the experiments described herein, and the relevant optical layout is shown in figure 5.17. In particular, for these experiments the harmonics are generated using an asymmetric two-color field that allows for the generation of even and odd harmonics. A schematic for how this is implemented, and the resulting XUV harmonic comb, is shown in figure 5.18. The harmonic comb that is generated in this way is more energetically dense, and it can be used to more finely sample the ground and excited states of the system. In particular, there can be light between the harmonics that enable the generation of pseudo-continuum of XUV light. This feature is important because it enables the observation of several LIS, as will be shown later.

Using the typical XUV harmonic source shown in figure 5.18, the ground state photoabsorption spectrum of the $3s3p^6np$ states can be measured. This is done by measuring the harmonic spectrum with and without argon gas in the gas cell that is shown in figure 5.17. From these two measurements, the optical density can be calculated as a function of photon energy ω using Beer-Lambert's Law as

$$\mathrm{OD}(\omega) = -\log(T(\omega)) = -\log\left(\frac{I_{\mathrm{on}}}{I_{\mathrm{off}}}\right) = \frac{\rho l}{\ln 10}\sigma(\omega), \tag{5.80}$$

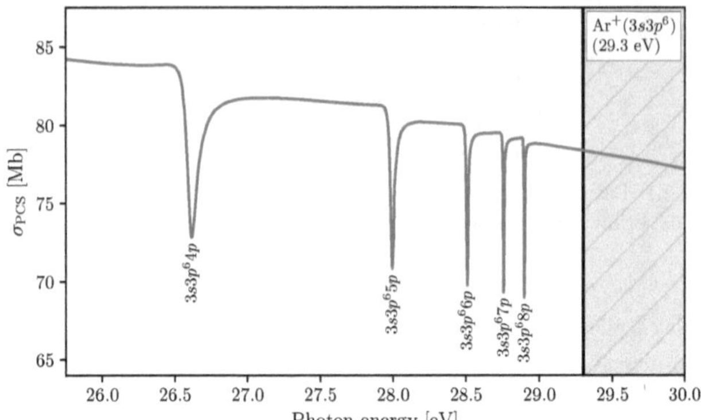

Figure 5.16: Photoabsorption cross section of the Argon $3s3p^6np$ Fano resonances (blue curve), with only resonances up to $n = 8$ shown. Grey shaded area indicates the energetic region above the $\mathrm{Ar}^+(3s3p^6)$ ionization threshold. Values used to calculate this cross section are shown in Table 5.1.

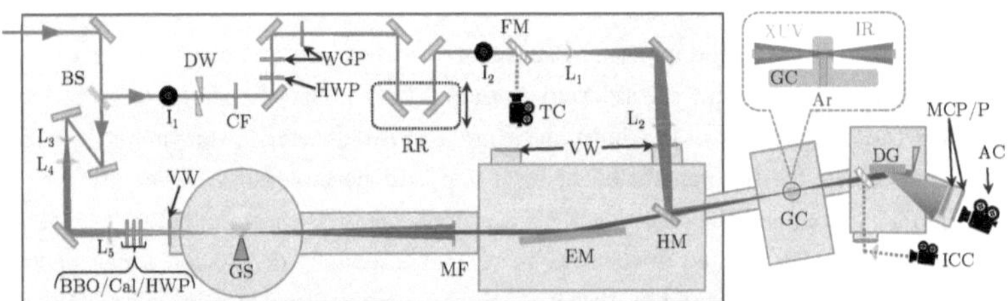

Figure 5.17: Schematic of the optical setup for the experiments described in this chapter. **BS**: Beamsplitter (Thorlabs BSF20-C), $\mathbf{I}_{1,2}$: Irises used for alignment. **DW**: Delay wedges for fine delay control. **CF**: Color filter (Thorlabs FELH1000). **HWP**: Half-wave plate. **WGP**: Wire grid polarizer. **RR**: Retro reflector for coarse delay adjustment. **FM**: Flip mirror. **TC**: Thermal camera used for alignment. $\mathbf{L}_1$: $f = -300$ mm lens (Thorlabs LF1015-C). $\mathbf{L}_2$: $f = 500$ mm lens (Thorlabs LA1380-C). **VW**: Vacuum window, 3 mm CaF$_2$, **HM**: Hole mirror with 10 mm hole. $\mathbf{L}_3$: $f = -400$ mm lens. $\mathbf{L}_4$: $f = 500$ mm lens. $\mathbf{L}_5$: $f = 400$ mm lens. **BBO**: Second-harmonic generation crystal. **Cal**: Calcite. **GS**: Gas source for HHG. **MF**: Aluminum filter. **EM**: Ellipsoidal mirror. **GC**: Gas cell. **RM**: Removable mirror for *in-situ* diagnotics. **ICC**: camera for *in-situ* diagnotics. **DG**: VLS diffraction grating. **BB**: Baffles to block zero order diffraction. **MCP/P**: Microchannel plate and phosphor. **AC**: Andor Neo 5.5 CMOS camera.

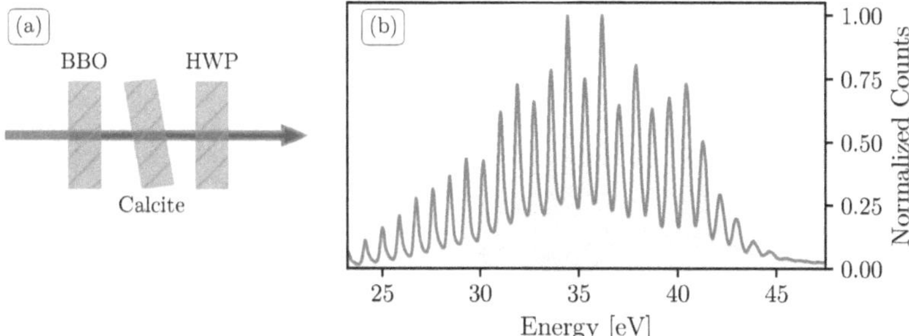

Figure 5.18: (a) Schematic of $\omega + 2\omega$ scheme to generate even and odd harmonics. BBO is used to generate 2ω. Calcite is used to temporally overlap the two colors at the gas source. The half wave plate is used to rotate the polarization of the ω field to match the 2ω field, thereby creating an asymmetric linearly polarized two-color field at the focus. (b) Typical harmonic spectrum used in the experiments described in the chapter. The harmonics are generated in an $\omega + 2\omega$ scheme where the fundamental wavelength is 1430 nm. The two-color field used to generate the harmonics has roughly equal intensity in the ω and 2ω components.

where $T(\omega)$ is the transmission, $I_{\text{on(off)}}$ is the spectrum with the gas on (off), ρ is the gas density in the interaction region, l is the length of the interaction region, and $\sigma(\omega)$ is the absorption cross section. For the gas cell used in these experiments, the length l is fixed by the cell design at 2 mm, and the density ρ is set by the backing pressure of the gas cell to achieve the desirable amount of absorption.

The ground state OD of argon measured using the harmonic spectrum in figure 5.18 (b) is shown in figure 5.19. From this measurement there are three clear features that stand out in the data. One is an overall decreasing absorption at higher photon energies that is consistent with the non-resonant background absorption that is expected. This is due to the fact that the absorption cross section decreases monotonically above the ionization potential when neglecting the effect of additional resonances [71, 120, 121]. The second clear feature is an oscillation in the measured OD that is most evident at photon energies above 33 eV. These oscillations are artifacts of the measurement that arise from one of several sources. The first possibility is a difference in background counts between the two cases that are being considered (gas on versus gas off). This can be seen by assuming that the measured spectral amplitudes $\tilde{I}_{\text{on,off}}(\omega)$ differ from the true spectral amplitude $I_{\text{on,off}}(\omega)$ by a constant amount that can be energy dependent,

$$\tilde{I}_{\text{on}}(\omega) = I_{\text{on}}(\omega) + a(\omega)$$
$$\tilde{I}_{\text{off}}(\omega) = I_{\text{off}}(\omega) + b(\omega). \tag{5.81}$$

This means that the measured transmission function $\tilde{T}(\omega)$ will differ from the true transmission function $T(\omega)$ by the relationship

$$\tilde{T}(\omega) = \frac{\tilde{I}_{\text{on}}}{\tilde{I}_{\text{off}}} = \frac{I_{\text{on}} + a}{I_{\text{off}} + b} \approx T(\omega)\left(1 - \frac{a(\omega)}{I_{\text{off}}(\omega)}\right) + \frac{b(\omega)}{I_{\text{off}}(\omega)}, \tag{5.82}$$

where a/I_{off} and b/I_{off} are assumed to be small and only terms up to $\mathcal{O}((1/I_{\text{off}})^2)$ are retained. Since this linear relationship between the measured and true transmission depends upon the harmonic spectral amplitude, one would expect to see modulations in the calculated OD that follow the spacing of the harmonic spectrum. Thus, for the harmonic spectrum that was used for these measurements that consists of even and odd harmonics, one would expect to see a modulation in the measured OD with a period corresponding to the photon energy of the fundamental wavelength used to generate the harmonic spectrum. This expected modulation at 0.87 eV for a fundamental wavelength of 1430 nm is exactly what is seen in the measured OD in figure 5.18. That being said, it is possible to make further assumptions about the stability of the harmonic spectrum in the time between I_{on} and I_{off} are measured that can also predict modulations in the measured OD that have a period of the fundamental wavelength. Regardless of the particular origin of these modulations, they are typically filtered out using some type of band pass frequency filter when analyzing the measured data [78, 122].

The third and final feature that is apparent in the measured OD shown in figure 5.19 are the $3s3p^6np$ Fano resonances of interest. In the measured OD, there is a clear dip in absorption at 26.6 eV with a series of dips that occur at higher energies, and these can be identified as the $4p$, $5p$, and $6p$ resonances.[13] This measured OD can be fit to the photoabsorption cross section calculated using

$$\text{OD}(\omega) = \frac{\rho l}{\ln 10}\left[\sigma_{\text{NR}}(\omega) + \sum_n \sigma_n W_n(\omega)\frac{(q_n + \epsilon_n)^2}{\epsilon_n^2 + 1}\right] \circledast \left[\frac{1}{\nu\sqrt{2\pi}}e^{-\frac{(\hbar\omega)^2}{2\nu^2}}\right] \tag{5.83}$$

where the sum is over the np states with resonance parameters given by q_n and $\epsilon_n = (\hbar\omega - E_{r,n})/(\Gamma_n/2)$, $\circledast$ represents convolution with the Gaussian of standard deviation ν that is chosen to represent the PSF of the spectrometer, and σ_n, ν, and ρl are fit parameters. The function $W_n(\omega)$ is a window function that is a convolution of a Gaussian and a boxcar, and this window function is used to limit the range of each resonance [90, 117]. In principal, the non-resonant background absorption could be approximated by the cross section from

---

[13]Unless stated otherwise, the $3s3p^6np$ will be referred to as the np resonance.

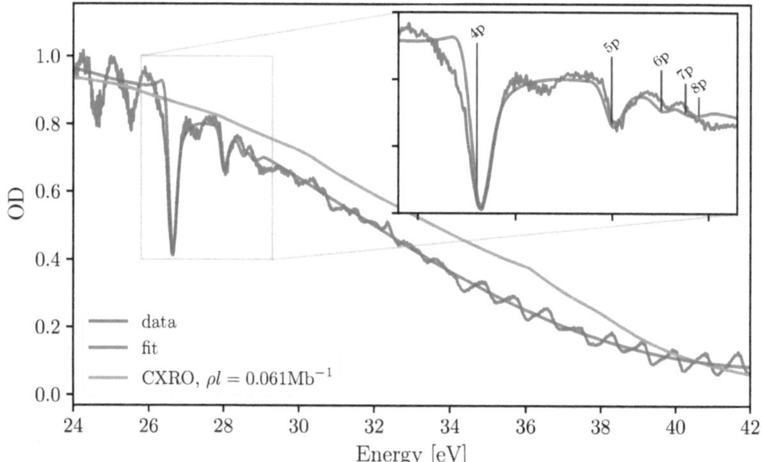

Figure 5.19: Measured ground state photoabsorption of $3s3p^6np$ Fano resonances. Blue curve is experimentally measured OD. Red curve is a fit to the experimental data using equation 5.83. Inset shows the resonance positions of the np resonances used in the calculation.

CXRO, however this does not agree well with the experimentally measured OD, as can be seen by the green curve in figure 5.19. Instead, the non-resonant absorption σ_{NR} is extracted from the measure OD by fitting to an expansion given by

$$\sigma_{\mathrm{NR}}(\omega) \approx a_0 + a_1(\hbar\omega - E_1) + a_2\mathrm{erfc}(w_2(\hbar\omega - E_2)), \tag{5.84}$$

where a_i, E_i, and w_2 are fit parameters and erfc is the complimentary error function. This choice of expansion was made for ease of fitting using a least-squares method, and it could easily be replaced by an expansion in a different basis.

The result of fitting equations 5.83 and 5.84 to the experimental data is shown as the red line in figure 5.19, and the inset in figure 5.19 shows the resonance energy positions of the $4p$ to $8p$ resonances. As can be seen, the $4p$ and $5p$ resonances show reasonably good agreement between the measured OD and the fit, however above the $5p$ resonance it is not possible to clearly identify any structure. This is likely due to modulations in the measured OD due to either fluctuations in the spectral amplitude of the harmonics or non-zero background counts, as described previously. That being said, the reasonable agreement between measurement and theory demonstrates the ability to resolve the np resonances, and by applying an appropriate dressing field, the dynamics described previously should be observed.

To that end, the dressing field parameters using the setup detailed in figure 5.17 were selected to enable a high dressing intensity given the available laser systems. This is achieved through the use of a telescope to expand the beam profile to minimize losses on the hole mirror that is used to recombine the XUV and the dressing field. For the experiments described below, the wavelength used is 1430 nm, and the pulse duration is nominally 60 fs. The range of peak intensities available for this experiment is 0.33 - 42 TW/cm^2, and this can be finely controlled through the use of a waveplate and wiregrid polarizer. Further details of the dressing setup are in section 2.4.2.

5.5.2 Intensity Dependence: Light-Induced Attenuation

Now that the theoretical background and experimental setup has been thoroughly established, attention can be turned to the experimental results. To begin, it is useful to first establish the intensity dependence of the features which are observed in the ATS spectrogram, and this will be measured at a fixed delay which corresponds to temporal overlap between the XUV and dressing pulses. Doing so will develop insight into the nature of resonant effects that are critical for this wavelength dressing field.

The measured intensity dependence of the dressed autoionizing states is shown in figure 5.20. The change in optical density ΔOD that is shown in this figure is given by

$$\Delta\text{OD} = -\log\left(\frac{I_{\text{on}}}{I_{\text{off}}}\right) \propto \sigma_{\text{on}} - \sigma_{\text{off}} \tag{5.85}$$

where $I_{\text{on(off)}}$ is the harmonic spectrum with (without) the presence of the dressing field, and it is related to the difference in photoabsorption cross section that the dressing field induces. As can be seen in figure 5.20, there are several distinct features that exhibit strong intensity dependence. It is expected to observe a change in the photoabsorption in the vicinity of the bright np resonances, however there are additional features that are observed between np states that are dipole allowed from the ground state. To elucidate this point, a level diagram consisting of both the bright np states and the dark $3s3p^6ns$ and $3s3p^3nd$ states is shown next to the measured ΔOD in figure 5.20, and the energies of these states are given in table 5.2 [117, 120, 121, 123–125]. Comparing the level diagram to the measured ΔOD, at the photon energy corresponding to the $4p$, $5p$, and $6p$ resonances there is a similar feature that can be observed. There is an increase in absorption (red on the color map of figure 5.20) that exhibits little shift in photon energy and an amplitude that quickly saturates with increasing peak intensity. These features will be attributed to suppression of the dipole amplitude from direct ionization of the np states. The other set

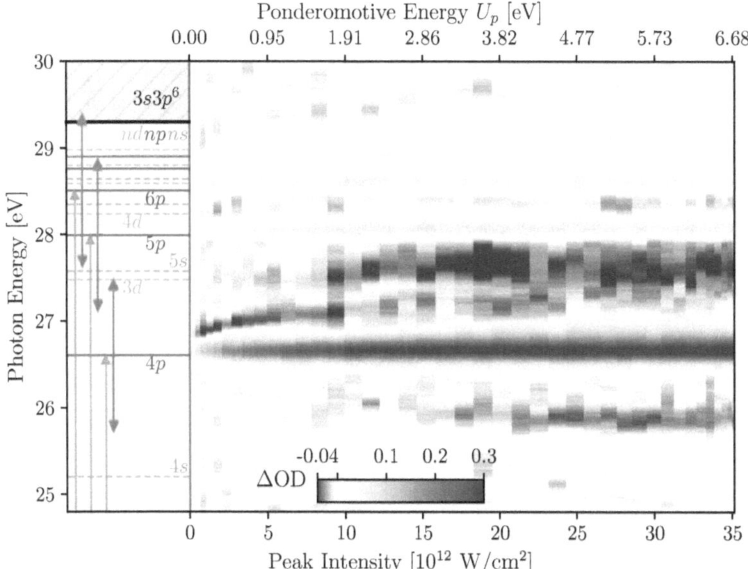

Figure 5.20: Intensity dependence of measured ΔOD near $3s3p^n l$ resonances in argon. Left portion shows level diagram with bright (dark) states given by solid (dashed) lines. Red arrows show dressing photon energy, and blue arrows show XUV transitions from ground state. Right portion of figure shows the measured ΔOD.

of features that can be seen in figure 5.20 occur at energies that do not correspond to any of the bright np states, and they generally consist of an increase in absorption (blue on the color map used). These will be shown to be light-induced states that naturally arise from the dressed atom picture using Floquet theory. The structure of these light-induced states are critically dependent upon the wavelength of the dressing field, and the wavelength in this case is near-resonant to transitions between bright and dark states.

To further expound upon this, we will first consider the features that arise at the photon energies corresponding to the np resonances. These increases in absorption can best be understood by utilizing the DCM and LIA theory that was established in sections 5.3 and 5.3.2. Generally speaking, the LIA model describes the suppression of the dipole induced by the XUV pulse through direct ionization of the excited state by the dressing pulse, and the DCM allows for this effect to be analytically calculated for a finite dressing pulse while

$3s3p^6nl$ excited states					
Level	E_r [eV]	Level	E_r [eV]	Level	E_r [eV]
				$3d(^1\mathrm{D})$	27.48
$4s(^1\mathrm{S})$	25.2	$4p(^1\mathrm{P})$	26.605	$4d(^1\mathrm{D})$	28.24
$5s(^1\mathrm{S})$	27.58	$5p(^1\mathrm{P})$	27.994	$5d(^1\mathrm{D})$	28.59
$6s(^1\mathrm{S})$	28.35	$6p(^1\mathrm{P})$	28.509	$6d(^1\mathrm{D})$	28.80
$7s(^1\mathrm{S})$	28.64	$7p(^1\mathrm{P})$	28.757	$7d(^1\mathrm{D})$	28.91
		$8p(^1\mathrm{P})$	28.898	$8d(^1\mathrm{D})$	28.98

Table 5.2: Energy of $3s3p^6nl$ states in Ar [117, 120, 121, 123–125].

assuming the XUV pulse is a δ-function in time. In section 5.3.2, it was established that the amplitude a_1 is related to the population of the excited state which can be directly ionized by the dressing field, and its relationship to the ionization probability and rate is given be equation 5.64. Using equation 5.67, the ΔOD will be related to the amplitude a_1 by

$$\Delta\mathrm{OD}(\omega,\tau) \propto \sigma_{\mathrm{on}}(\omega,\tau) - \sigma_{\mathrm{off}}(\omega,\tau)$$
$$\propto \mathrm{Im}\left[i\gamma(q-i)^2 \int_0^\infty \left(a_1(t,\tau) - 1\right)e^{-i\delta t}e^{-\gamma t}\,\mathrm{d}t \right]. \tag{5.86}$$

The effect of a_1 on the absorption spectrum can be intuitively understood by looking at limiting cases of this equation. For the case of $a_1 \approx 1$ when the field is not intense enough to significantly ionize, $\Delta\mathrm{OD} \approx 0$ and there is no change in the absorption spectrum as expected. For the other limiting case of $a_1 \approx 0$, then $\sigma_{\mathrm{on}} \approx 0$ and $\Delta\mathrm{OD} \propto -\sigma_{\mathrm{off}}$. This implies that the effect of a_1 is to reduce the contribution that a particular resonance makes to the total cross section, and this provides an intuitive explanation for the features that arise in the at photon energies corresponding to the np resonances in the intensity spectrogram in figure 5.20. Specifically, they represent depletion of the np resonances by the dressing pulse until a saturation intensity is reached.

This qualitative understanding can be made more quantitative by calculating equation 5.86 for representative XUV and IR dressing pulses. The IR dressing pulse that will be used in this calculation is shown in figure 5.21, and the XUV pulse is taken to be a pulse of 150 as in duration. This pulse duration is much smaller that the expected XUV APT duration of 60 fs, however, it is a reasonable representation of each pulse in the APT [14, 39, 126, 127]. The result of calculating the amplitude a_1 for the $4p$ resonance is shown in figure 5.22. In figure 5.22 (a), the dipole in the time domain is calculated on resonance ($\delta = 0$) for the range of intensities used in the experiment, and the ionization rate is calculated using the PPT model. As can be seen, as the intensity increases the duration of dipole decreases precipitously until it is limited by the duration of the XUV pulse. The effect that this has on the cross section in the presence of the dressing field is shown in figure 5.22 (b), and it

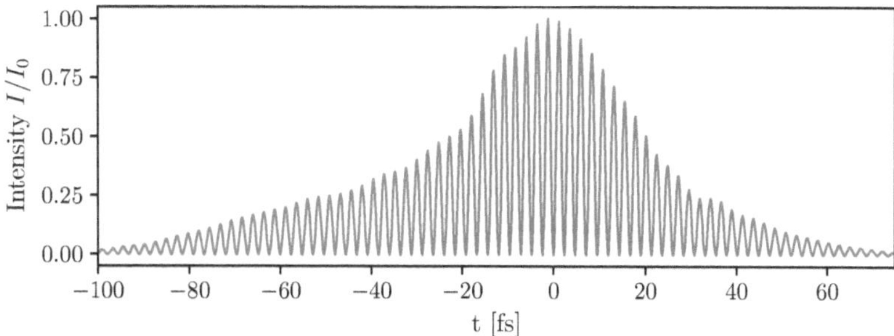

Figure 5.21: IR dressing pulse intensity used for calculations, and is the result of a FROG measurement of the output of the TOPAS at the wavelength used for these experiments (1430 nm). Normalized by the peak intensity that ranges from 0.2 - 35.3 TW/cm^2 in the DCM/LIA calculations.

is eminently evident that as the intensity increases the Fano profile of the ground state is suppressed until the cross section is a constant proportional to the background cross section. This aligns with the intuitive description presented previously.

This calculation can be extended to the $5p$ and $6p$ resonances to calculate the total ΔOD. This is done by treating each state independently and assuming that the total ΔOD is a sum of the individual contributions from each resonance,

$$\Delta\mathrm{OD}_{tot}(\omega) = \left[\sum_n A_n W_n(\omega)\big(\sigma_{\mathrm{on},n}(\omega) - \sigma_{\mathrm{off},n}(\omega)\big) \right] \circledast \left[\frac{1}{\nu\sqrt{2\pi}} e^{-\frac{(\hbar\omega)^2}{2\nu^2}} \right] \tag{5.87}$$

where A_n is an amplitude that is used to fit to the measured ΔOD, $W_n(\omega)$ is a window function that limits the range of the calculated σ_n to $\pm 3\Gamma_n$, and $\circledast$ represents convolution with the Gaussian of standard deviation $\nu = 0.08$ eV that is chosen to represent the PSF of the spectrometer. The window function is applied because the form of the dipole used in the DCM inherently assumes a flat continuum, and this approximation is only valid near the resonance energy [90]. The resulting calculation is shown in figure 5.22 (b), and this includes averaging over the range of IR intensities within the interaction focal volume for a given peak intensity of the dressing field. Generally speaking, the agreement with the measured ΔOD is quite good near the energies corresponding to the np resonances, however the features between each np resonance are not reproduced, as is expected. To examine the agreement further, the maximum ΔOD near each resonance is shown as a function of peak intensity in figures 5.23 (c) - (e) for both the measured and calculated ΔOD. The amplitude A_n in equation 5.87 is a fit parameter that is used to scale the calculated ΔOD by a constant

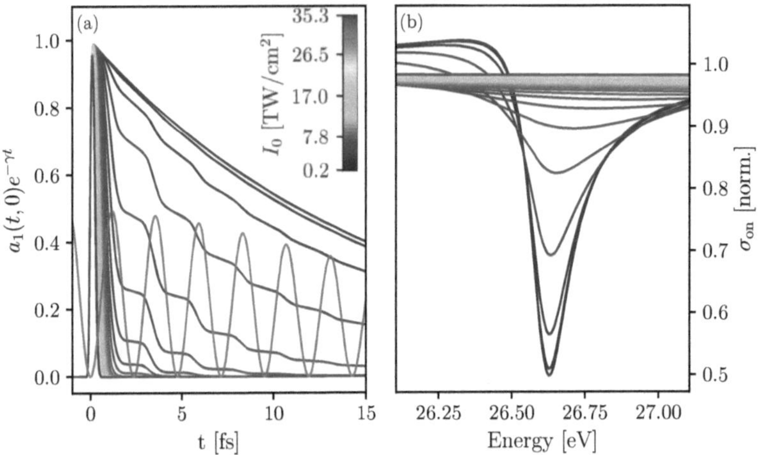

Figure 5.22: (a) Dipole amplitude a_1 in the time domain as a function of the peak intensity of the IR dressing pulse (intensities given by color bar and dressing pulse shown in red). (b) Corresponding cross section in the presence of the dressing pulse, and colors correspond to the same dressing peak intensity in (a).

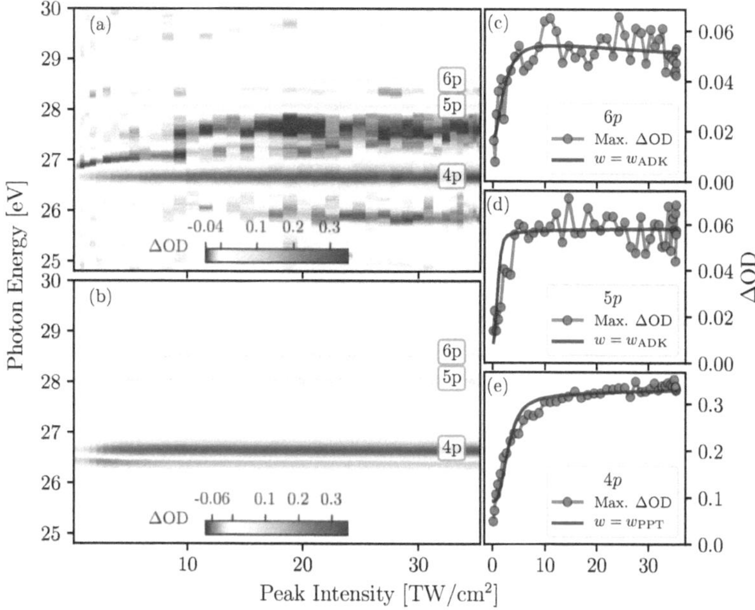

Figure 5.23: (a) Measured ΔOD as function of peak dressing intensity with labels for the np resonances that will be used in the DCM calculation for comparison. (b) Calculated ΔOD as a function of intensity. (c)-(e) Comparison of maximum ΔOD at each np resonance position. Ionization rate is calculated using PPT for the $4p$ state (e) and using ADK for the $5p$ (d) and $6p$ (c) state.

for each resonance, and after fitting for these parameters, we can see that the intensity scaling of the maximum ΔOD is well reproduced by the DCM/LIA calculation using the PPT ($4p$) and ADK ($5p$ and $6p$) models. Essentially, this suggests that the features in the intensity spectrogram that are at each resonance energy position correspond to ionization of the discrete state. Further improvements can be made to the agreement between the calculation and the measured ΔOD by more accurately calculating the ionization of these weakly bound excited states and by including the full APT in the calculation.

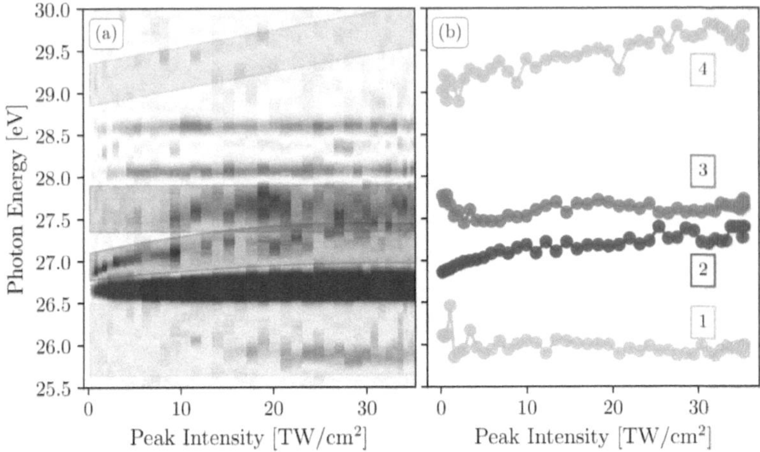

Figure 5.24: (a) Measured ΔOD as function of peak dressing intensity with labels for the np resonances that will be used in the DCM calculation for comparison. (b) Calculated ΔOD as a function of intensity. (c)-(e) Comparison of maximum ΔOD at each np resonance position. Ionization rate is calculated using PPT for the $4p$ state (e) and using ADK for the $5p$ (d) and $6p$ (c) state.

5.5.3 Intensity Dependence: Light-Induced States

In the previous section, the origin of several of the intensity dependent features found in the ΔOD intensity spectrogram was explained in the context of the DCM/LIA model, however this description was only able to account for the features observed at the np resonance energy positions. This section will identify the other features as light-induced states (LIS) that arise from the dressed atom picture, but first it is important to identify each of the LIS in the intensity spectrogram. This identification is shown in figure 5.24 and consists of choosing an energetic window as a function of intensity to find the position of each LIS in energy as a function of intensity by fitting the ΔOD to a lineshape. In principle, each of these states can couple to the nearby continua, and this will result in an asymmetric Fano lineshape that depends upon the coupling strength between each state and the continuum. This can be seen in the LIS 4 (turquoise), however the other LIS have a symmetric shape. Regardless of their shape, each LIS energy position is extracted as a function of intensity, and this will allow for further analysis.

To explain the light-induced features that occur away from the np resonance positions in the measured intensity spectrogram in figure 5.20, we must go beyond the DCM/LIA model and include the presence of nearby states to account for the LIS, see section 5.5.3 and 5.4. As can be seen in the level diagram in figure 5.20, there are many states within the vicinity of the np resonances that don't appear in the ground state photoabsorption spectrum because they are not dipole allowed from the ground state. That being said, they play a critical role in the observed features because many of these states are dipole allowed from the np states and some of these transitions are actually near-resonant. In fact, for the dressing photon energy of 0.867 eV that was used, each np resonance has a nearly resonant transition to one of the dark states. The detunings for three of these transitions are $\Delta_{4p \to 3d} = -0.008$ eV, $\Delta_{5p \to 7d} = -0.049$ eV, and $\Delta_{6p \to 5s} = -0.062$ eV. The $4p \to 3d$ transition in particular is within the roughly 0.06 eV bandwidth of the dressing pulse. This near-resonant condition implies that the Raman-like interpretation of light-induced states as a two-photon process involving absorption of one XUV photon and the absorption or emission of an IR photon to reach a dark state doe not apply [97, 112]. In this case, it is most appropriate interpret the light-induced states as arising naturally from dressed states of the atom in a strong dressing field, and this can be understood in terms of Floquet theory that was introduced in section 5.4.

In section 5.4, the adiabatic Floquet theory was introduced and calculated for a simple two level system consisting of a bright and dark state dressed by a field with photon energy nearly resonant to the transition between the two states. This simple example can be used in the case of the $4p$ state to get an understanding of the basic effects that are occurring. In the adiabatic Floquet theory, the time evolution of the Floquet states is a trivial phase evolution, however the challenge lies in calculating the Floquet states and their energies. This is done by diagonalizing the Floquet Hamiltonian $\hat{H}_F$ given by equation 5.73, and this takes the matrix form given by equation 5.74 when considering a two-level system and up to $n = 2$. The result of this calculation for the $4p$ bright state and the $3d$ dark state is shown in figure 5.25 for an assumed dipole matrix element of $\mu = 1.987$ a.u. In this figure, the Floquet ladder of quasienergies is plotted on top of the measured ΔOD intensity spectrogram for comparison. Due to the resonant nature of this case, there is a splitting that occurs, and this is the well-known Autler-Townes effect which gives rise to a symmetrically split doublet. [97, 105, 128]. This splitting about the bright $4p$ state can be seen to overlap with two of the LIS just above and below the $4p$ state. This gives a qualitative understanding of the origin of two of the LIS as arising from the Autler-Townes doublet that is formed by resonantly coupling the $4p$ to the $3d$ state.

The simple two-level system is excellent for gaining an intuitive understanding of the effects that are observed, however it is insufficient for quantitative agreement with all of the LIS that can be seen in the intensity spectrogram. To this end, we need to include more states in the Floquet calculation. Specifically, all states np, nd, and ns states up to

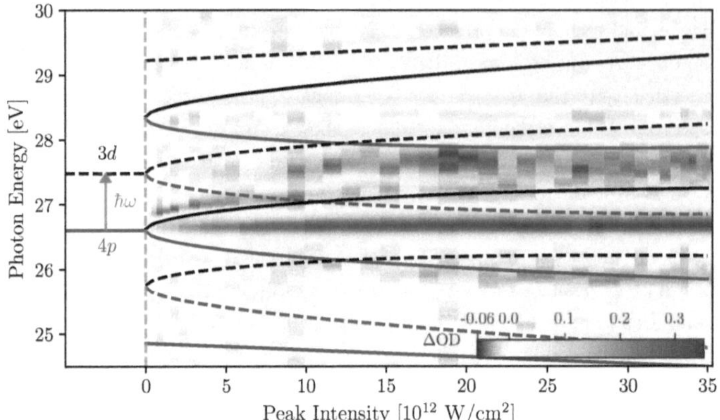

Figure 5.25: Floquet dressed states as a function intensity calculated by a two state adiabatic Floquet theory. States with solid lines are dipole allowed from the ground state, and states with dashed lines are not dipole allowed. The pronounced Autler-Townes splitting of the $4p$ state corresponds with two of the LIS in the intensity spectrogram (1 and 2 in figure 5.24).

$n = 6$ will be included, a total of 10 states. This is done because the Floquet ladders from each state strongly interact through avoided crossings, and this has a profound effect on the intensity dependence of the quasienergy spectrum. Including these additional states means that the matrix representing the Floquet Hamiltonian in equation 5.74 needs to be generalized to include all of these states. The basic form of the Hamiltonian is still the same block tri-diagonal as it was in equation 5.74, however it can now be written as

$$\hat{H}_F = \begin{bmatrix} \hat{H}_0 - 2\omega\hat{I} & \hat{H}_1 & \hat{0} & \hat{0} & \hat{0} \\ \hat{H}_{-1} & \hat{H}_0 - \omega\hat{I} & \hat{H}_1 & \hat{0} & \hat{0} \\ \hat{0} & \hat{H}_{-1} & \hat{H}_0 & \hat{H}_1 & \hat{0} \\ \hat{0} & \hat{0} & \hat{H}_{-1} & \hat{H}_0 + \omega\hat{I} & \hat{H}_1 \\ \hat{0} & \hat{0} & \hat{0} & \hat{H}_{-1} & \hat{H}_0 + 2\omega\hat{I} \end{bmatrix}, \qquad (5.88)$$

where the choice is made to include only up to two photon transitions ($n = 2$) and each element is a 10x10 matrix. When represented in the $4p, 3d, 4d, 4s, 5s, 5d, 6s, 5p, 6p, 6d$ basis, the matrix H_0 takes the form

$$\hat{H}_0 \rightarrow I_{10}[\omega_{4p}, \omega_{3d}, \omega_{4d}, \omega_{4s}, \omega_{5s}, \omega_{5d}, \omega_{6s}, \omega_{5p}, \omega_{6p}, \omega_{6d}]^T \qquad (5.89)$$

where I_{10} is the 10x10 identity matrix and ω_{nl} are energies of each nl resonance. Additionally, in the same basis the matrix $H_{\pm 1}$ takes the form

$$\hat{H}_{\pm 1} \rightarrow \begin{bmatrix}
0 & \Omega_0/2 & \Omega_1/2 & \Omega_2/2 & \Omega_3/2 & \Omega_4/2 & \Omega_5/2 & 0 & 0 & \Omega_{18}/2 \\
\Omega_0/2 & 0 & 0 & 0 & 0 & 0 & 0 & \Omega_6/2 & \Omega_{12}/2 & 0 \\
\Omega_1/2 & 0 & 0 & 0 & 0 & 0 & 0 & \Omega_7/2 & \Omega_{13}/2 & 0 \\
\Omega_2/2 & 0 & 0 & 0 & 0 & 0 & 0 & \Omega_8/2 & \Omega_{14}/2 & 0 \\
\Omega_3/2 & 0 & 0 & 0 & 0 & 0 & 0 & \Omega_9/2 & \Omega_{15}/2 & 0 \\
\Omega_4/2 & 0 & 0 & 0 & 0 & 0 & 0 & \Omega_{10}/2 & \Omega_{16}/2 & 0 \\
\Omega_5/2 & 0 & 0 & 0 & 0 & 0 & 0 & \Omega_{11}/2 & \Omega_{17}/2 & 0 \\
0 & \Omega_6/2 & \Omega_7/2 & \Omega_8/2 & \Omega_9/2 & \Omega_{10}/2 & \Omega_{11}/2 & 0 & 0 & \Omega_{19}/2 \\
0 & \Omega_{12}/2 & \Omega_{13}/2 & \Omega_{14}/2 & \Omega_{15}/2 & \Omega_{16}/2 & \Omega_{17}/2 & 0 & 0 & \Omega_{20}/2 \\
\Omega_{18}/2 & 0 & 0 & 0 & 0 & 0 & 0 & \Omega_{19}/2 & \Omega_{20}/2 & 0
\end{bmatrix} .$$

$$(5.90)$$

where $\Omega_i = \mu_i \mathcal{E}_{IR}$ are the Rabi frequencies of each transition and μ_i is the corresponding dipole matrix element. In principle, the dipole matrix elements are known and can be calculated [129]. However, the only matrix element directly found in the literature is $\mu_1 = \langle 4p|z|4d \rangle = 1.54$ a.u. [77, 130]. In lieu of directly calculating the matrix elements, their values can be obtained by fitting each of the Rabi frequencies Ω_i to the measured energy positions of the LIS seen in the experiment. This is done by using the known matrix element to generate a set of matrix elements that can be used as a reasonable ansatz, and then the dressed state spectrum is fit the to LIS energy position using a non-linear least-squares optimization routine where the matrix elements are the optimized parameters. The optimized matrix elements from this method are given in table 5.3, and the corresponding Floquet dressed state spectrum as a function of intensity is shown in figure 5.26. As can be seen by the comparison between the Floquet dressed states energy and the LIS energy in figure 5.26 (b), there is reasonably good agreement between the theory and the experimentally measured energies, especially LIS 1 and LIS 2 on either side of the $4p$ resonance. The state labels that are given in the legend of figure 5.26 (b) correspond to the basis states at zero field and are used only as a reference because in this case the Floquet states consist of strongly mixed basis states.

The reasonable agreement between the Floquet theory and the measured LIS energies demonstrates that the LIS can be interpreted as arising from a probing of the dressed atom. In principle, this calculation can be taken even further, and the ΔOD can be calculated from the Floquet dipole given by equation 5.78. However, this calculation should be performed with the actual dipole matrix elements to better examine the differences between the experiment and theory in amplitude and position if the light-induced features. That being said, the extracted matrix elements and corresponding Floquet theory can serve as a guide for analyzing the delay-dependent ΔOD in section 5.5.4.

$\langle np|z|nl\rangle$	3d	4d	4s	5s	5d	6s	6d
4p	1.505	1.54	3.276	1.62	0.112	0.592	1.148
5p	1.579	0.474	-0.113	1.366	0.057	1.006	1.039
6p	-0.961	0.105	-0.676	1.713	0.04	1.119	1.097

Table 5.3: Dipole matrix elements in atomic units that were the result of a fit between Floquet theory calculation and LIS energies. The matrix element $\langle 4p|z|4d\rangle$=1.54 a.u. is known from literature [130].

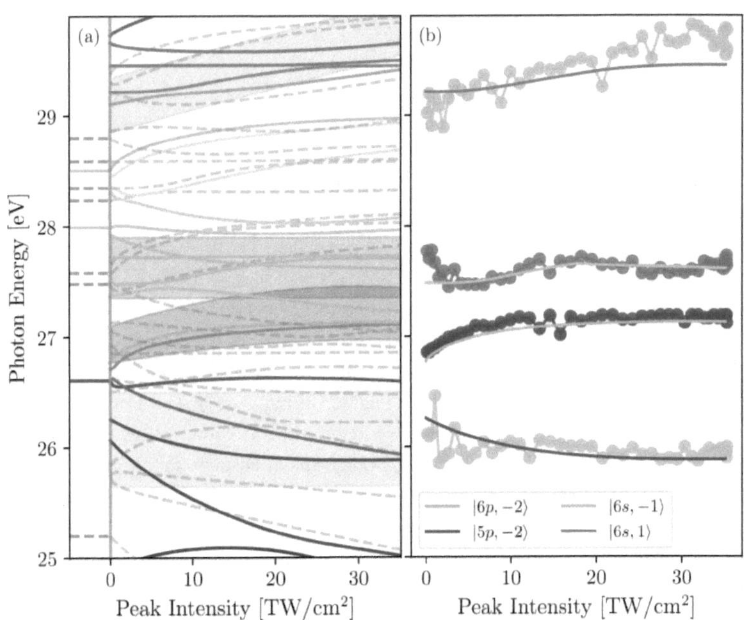

Figure 5.26: (a) Floquet dressed state spectrum calculated as a function of peak intensity. States that are not dipole allowed from the ground state are shown as grey dashed lines. States involved in the calculation are shown below zero intensity. (b) Comparison between the energy position of the LIS and the dressed states that best fit them. The residual deviation between the LIS energy position and the dressed states is used to determine the dipole matrix elements that are used in the Floquet theory calculation.

5.5.4 Delay Dependence

Now that the intensity dependent features have been fully explored, we now turn our attention to the delay dependent ΔOD. This measurement is performed at a series of fixed intensities and the delay between the dressing pulse and the XUV pulse is scanned through temporal overlap between the two pulses. This delay range covers roughly 200 fs because the duration of both the IR and XUV pulses is roughly 60 fs. At each delay a pair of spectra are taken consecutively with and without the dressing pulse. Each spectra requires at least 2 seconds to acquire, so the total experimental acquisition time is on the order of 3 hours per delay scan when the time delay steps need to be small enough to be able to resolve sub-cycle changes in the ΔOD. For all of the datasets shown in this section the time delay step size is 200 as. This step size was chosen to sufficiently resolve a signal at 2ω of the dressing field which has a period of 4.77 fs. It is important to be able to resolve sub-cycle signals because many processes such as strong-field ionization exhibit strong sub-cycle behavior because of their inherent intensity dependence, and in many ATS experiments a sub-cycle modulation of the photoabsorption is observed at 2ω due to one of several effects [80, 92, 97, 101, 113, 115, 131].

These delay measurements are shown in figure 5.27 (a) - (d) for peak intensities of 1.9, 10.9, 21.3, and 33 TW/cm^2, and it is obvious that the intensity dependent features that were discussed in sections 5.5.2 and 5.5.3 exhibit a strong delay dependence, as expected. In particular, at the np energies there is strong increase in absorption (red on the color map) that was previously described in terms of the DCM/LIA model as an attenuation of the dipole by direct ionization of the excited state. Qualitatively, we would expect this effect to strongest at temporal overlap between the two pulses and to quickly saturate in ΔOD amplitude, as shown in figure 5.23. To demonstrate this, line-outs are taken from the full spectrogram at the maximum ΔOD near each np resonance, and after taking a low-pass filter to remove higher frequency modulations, the result is shown in figure 5.28. These line-outs show the qualitative behavior that is expected with a quick saturation of the maximum ΔOD at temporal overlap between the two pulses (delay $\tau = 0$). Further calculations can be performed to provide more quantitative agreement between these line-outs and the DCM/LIA theory calculated for an XUV APT, however the qualitative agreement between the measured behavior and the previously calculated intensity dependence suggests that these features can be simply explained by applying PPT and ADK ionization models to account for the attenuation of the dipole.

To further examine the features that are present in the measured spectrograms, it is useful to examine the frequency components that contribute the ΔOD signal as a function of delay. This is done by simply Fourier transforming the ΔOD along the delay axis for every measured photon energy. The resulting two-dimensional absorption spectrum (2DAS) is shown in figures 5.27 (e) - (h). From these 2DAS, it is apparent that there is a separation

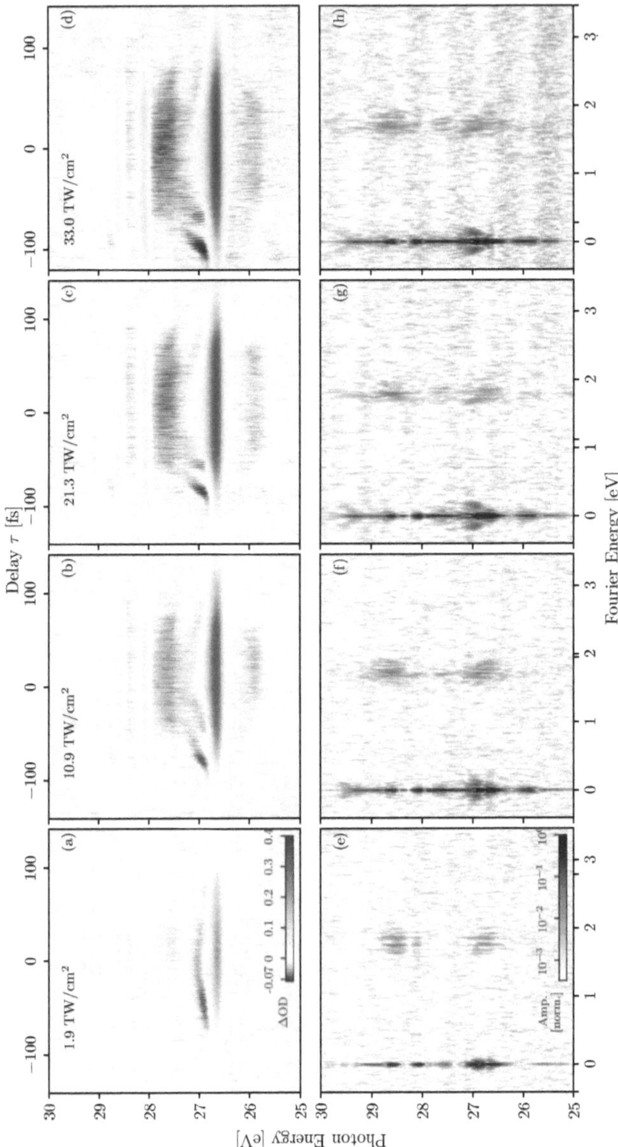

Figure 5.27: (a)-(d) Measured $\Delta\text{OD}(\tau)$ for different peak intensities as indicated in each plot. (e)-(h) Amplitude of the Fourier transform along the delay axis, the 2DAS. Shows clear 2ω oscillations and additional non-resonant structure at low Fourier energies. Negative frequencies are only shown to make features near zero Fourier energy more apparent.

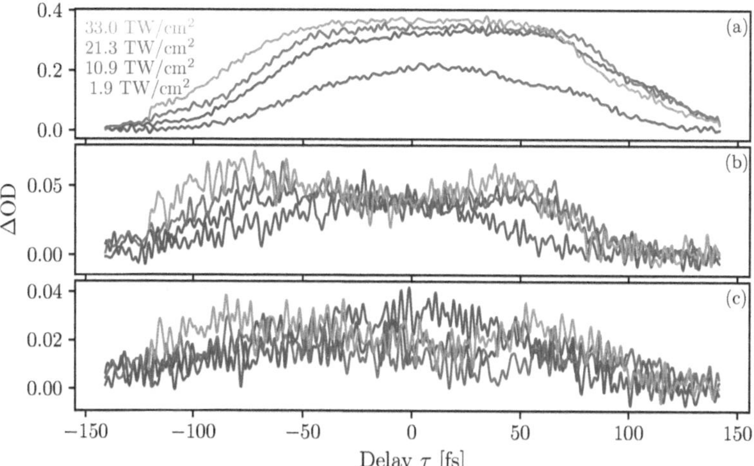

Figure 5.28: Line-outs of maximum ΔOD as function of delay for the $4p$ (a), $5p$ (a), and $6p$ (a). A low-pass Fourier filter was applied along the delay axis to remove 2ω oscillations for clarity.

of two sets of features in Fourier energy, and this is similar to the features seen in the calculations performed in section 5.3. The first set of features to focus on are those centered at 1.7 eV and correspond to a modulation of the ΔOD at a frequency of 2ω. At every intensity that was measured, the strongest Fourier amplitude occurs at the $4p$ and $6p$ energies, however there is still a small 2ω Fourier amplitude at the $5p$ resonance. There is also a pronounced splitting around the 2ω peak with a separation of 0.15 eV that is most apparent at the lowest intensity that was measured. There are several possible origins for these types of features and the two that will be examined herein are modulation due to ionization and wave-packet coupling by the dressing field [92].

To distinguish between these sources of sub-cycle modulations, a band-pass Fourier filter can be applied along the delay axis to isolate these frequency contributions. The Fourier band that was used is shown in figures 5.29 (e) - (h), and line-outs of the filtered ΔOD along the $4p$ and $6p$ energies is shown figures 5.29 (a) - (d) for each intensity measured. The $6p$ line-out has been arbitrarily shifted by a constant amount for clarity, and the inset in each plot shows the two line-outs over a small delay range without any shift applied. One possible source for these sub-cycle modulations is simply attenuation of the dipole due to ionization of the excited state, as is the case in the LIA model. This will lead to oscillations in the ΔOD at 2ω because the ionization probability is highest in a narrow temporal window at the peak of the field. So, when the delay between the APT and the dressing pulse is such that each pulse in the train is at a peak (valley) in the intensity profile, then the ΔOD signal is be weaker (stronger). This leads to an oscillation with period determined by the periodicity of the ionization probability, namely 2ω. In this case where the oscillations are driven by ionization, it would be expected that the ΔOD oscillation would be in phase across all of the np states. This is due to the fact that the oscillation period and phase is determined by the dressing pulse. This can be seen in the inset line-outs for the lowest intensity case shown in figure 5.27 (a) where the ΔOD for both the $4p$ and $6p$ are oscillating in phase. This is suggestive that the mechanism at play here is simply ionization by the dressing field. However, this is clearly not the case for higher intensities where the ΔOD is completely out-of-phase between the two states as a function of delay.

To explain the phase shift in the ΔOD between the $4p$ and the $6p$ states, we must consider the Fourier amplitude as a function of photon energy. In discussing the resonant cases handled by the DCM, see section 5.3, it was shown that for the case of a resonant coupling between two bright states separated by $\delta\omega$, there is a corresponding oscillation in the cross section an a frequency of $\delta\omega$. Furthermore, in the 2DAS, there is conveniently a line that effectively points between the two resonances being coupled, see figure 5.10 and 5.11, because it is given by the relationship $\delta = \pm\nu + \Delta\omega$. These lines are plotted on top of the 2DAS shown in figures 5.27 (f) - (h) for the $4p$ and $6p$ resonances assuming that they are the states being coupled. The signal is noisy, however there is reasonable agreement between

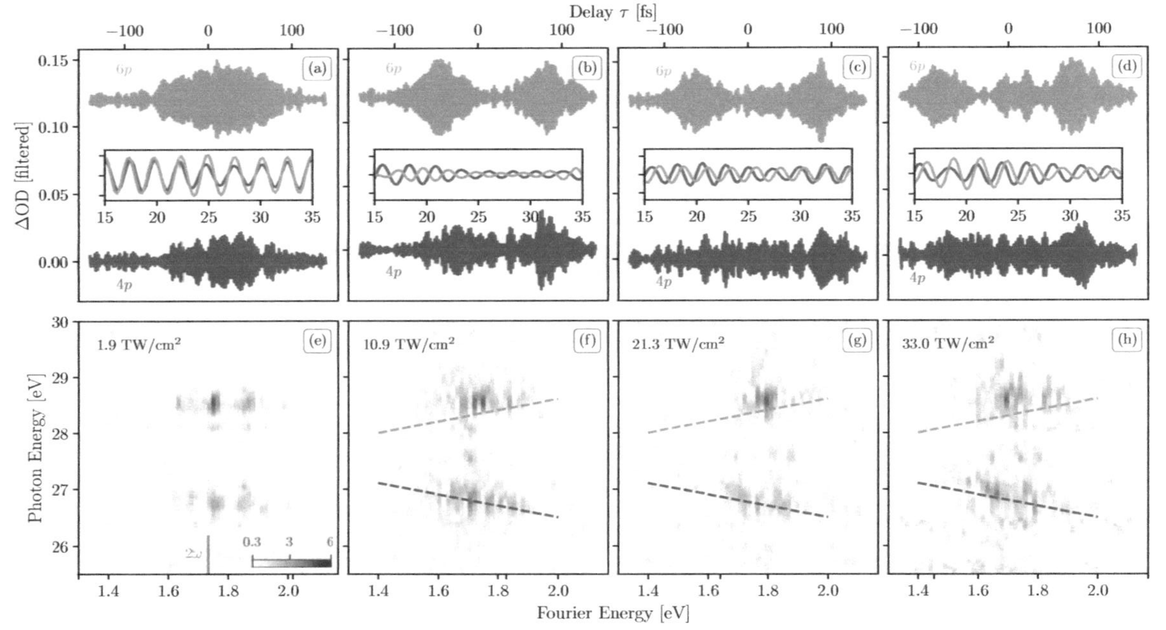

Figure 5.29: (a)-(d) Maximum of measured $\Delta\text{OD}(\tau)$ after applying a band-pass Fourier filter for $4p$ (purple) and $6p$ (blue). The $6p$ line-out is shifted for clarity. Insets shows the both line-outs for both states over a small delay window. (e)-(h) Fourier energy region used for the band-pass filter, and it is centered on 2ω. Red line show the 2ω energy, and the dashed lines represent coupling between $4p$ and $6p$ states.

the plotted lines and the amplitude, particularly for the intensity 10.9 TW/^{2} shown in figure
5.27 (f). Coupling between these two states can be understood in terms of coupling states
within a wave-packet. Initially, each pulse in the APT coherently excites a wave packet
consisting of the np resonances, and the dressing pulse acts to couple the $4p$ and $6p$ states
through a two-photon transition. This effect has been observed previously in helium, and
it was shown that this effect appears even in the the case of an APT instead of an IAP
[78, 92]. This is essentially due to the fact that the spacing of the pulses in the APT is close
to the period of the wave packet. For the $4p$ and $6p$ states their energy separation is 1.875
eV, and this is close to the 2ω energy of 1.734 eV. This notion of wave packet coupling
between the $4p$ and $6p$ states also helps to explain why the Fourier amplitude is larger for
both the $4p$ and the $6p$ than the $5p$.

The interplay between ionization and wave packet coupling can be used to explain the
sub-cycle modulations of the ΔOD at each resonance energy, however further interpretation
is needed to explain the non-resonant features that appear in the 2DAS at Fourier frequencies near zero. To show their effect on the ΔOD, a low pass Fourier filter is applied to
the measured ΔOD, and this is shown in figures 5.30 (a) - (d) along with the 2DAS near
zero Fourier frequency in figures 5.30 (e) - (h). As can be seen in these figures, the LIS
that were described in section 5.5.3 appear as a delay dependent feature, as is expected.
To provide qualitative understanding of the delay dependence of these LIS, the Floquet
dressed state spectrum can be calculated as a function of delay where the intensity of each
delay point is taken to be given by the autocorrelation of the dressing field shown in figure
5.21. The resulting Floquet dressed state spectrum is shown for the four peak intensity
used in the experiment in figures 5.31 (a) - (d), and the dressed states that corresponded to
the LIS observed in the intensity spectrogram are plotted on top of the ΔOD spectrograms
in figures 5.31 (e) - (h). The agreement between the observed features and the calculated
Floquet dressed state spectrum demonstrates that these non-resonant features in the 2DAS
can be interpreted as a probing of the dressed atom. Better quantitative agreement can be
made between the theory and the experiment with improved dipole matrix elements, and a
more thorough calculation of the Floquet dipole given by equation 5.78.

5.6 Conclusion

In this chapter, the theoretical background of Fano resonances as formulated by Ugo Fano
was established, and it was further extended into a time-domain interpretation where the
initial excitation was assumed to be infinitesimally short in time. This analytical form of
the dipole in the time domain allows for an intuitive understanding of the lineshape as the
phase of the dipole in the time-domain. Furthermore, it was established how an IR dressing
field can modulate this dipole as a function of delay between the initial excitation and the

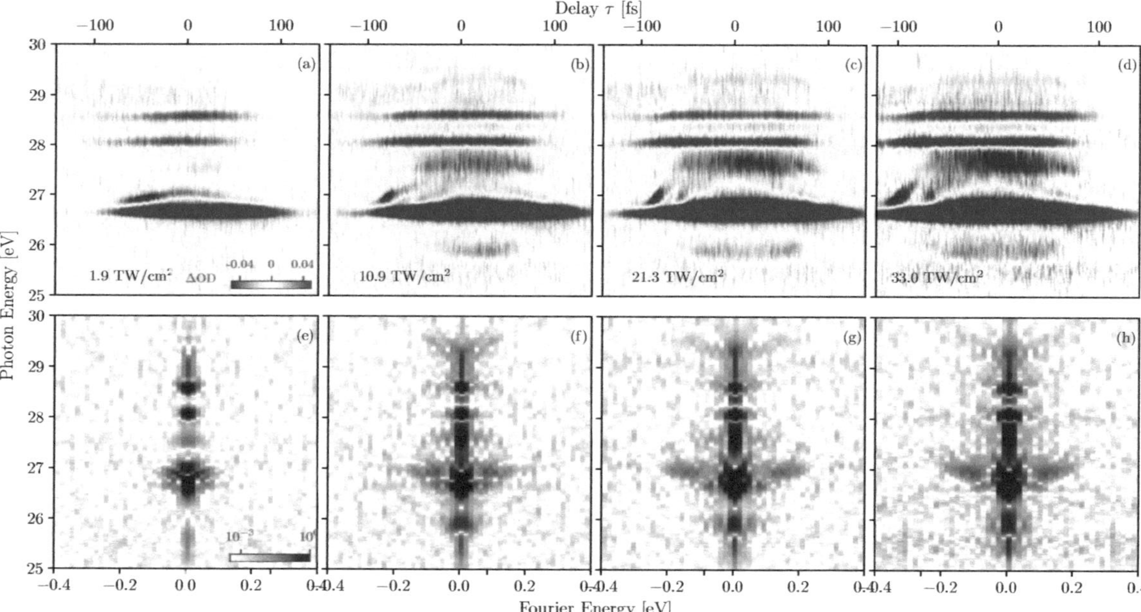

Figure 5.30: (a)-(d) Measured $\Delta OD(\tau)$ after applying a low-pass Fourier filter. (e)-(h) 2DAS highlighting the structures that are apparent at low Fourier energies.

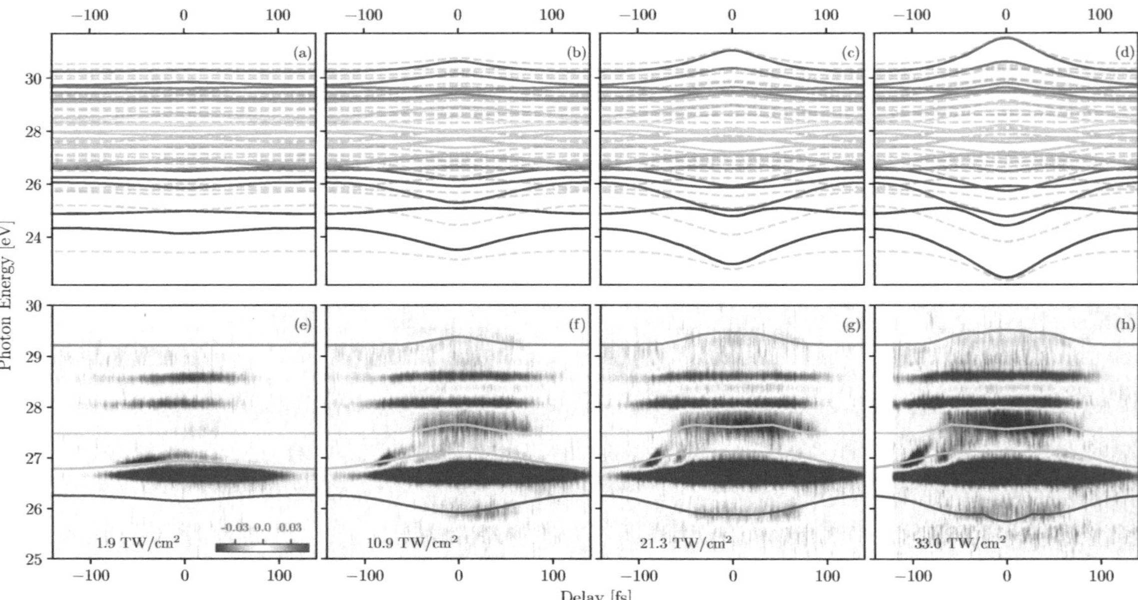

Figure 5.31: (a)-(d) Floquet dressed state spectrum calculated at each delay based upon the IR field in figure 5.21 and at the peak intensities shown below. (e)-(h) Measured $\Delta OD(\tau)$ after applying a low-pass Fourier filter with Floquet dressed state spectrum show for reference.

dressing pulse. This dipole control model was then extended to finite dressing pulses to account for the typical pulses used in the experiments presented herein. This model allows for an understanding of the features that are seen in the experimental data that appear close to each of the Fano resonance energies. An additional theory was established to account for light-induced states that cannot be explained by the dipole control model, and this model consists of an adiabatic Floquet theory that was initially developed for a simple two-state system to explain the Autler-Townes splitting that is seen in the experimental data. This model was extended to include more bright and dark states to account for further structures that are seen in the experimental data. The culmination of all of these theoretical and experimental efforts is an intuitive understanding of all of the features that are seen in the intensity- and delay-dependent measurements. This foundational knowledge of the inherent dynamics will be used in chapter 6 where the ultimate goal of measuring the real and imaginary part of the refractive index in a dynamic system is realized.

Chapter 6
COMPLEX ATTOSECOND TRANSIENT-ABSORPTION SPECTROSCOPY

6.1 Introduction

As seen in Chapter 5, a rich amount of information can be extracted from ATS experiments. Specifically, dynamics such as light-induced states and strong-field ionization of excited states induced by a dressing field can be deduced from the change in photoabsorption cross section. However, these experiments are limited by the fact that they only have access to the imaginary part of the complex refractive index of the medium of interest. There should also be a corresponding change in the real part of the complex refractive index that remains unobserved. In this Chapter, the techniques introduced in Chapters 3 and 4 are extended to measure both parts of the complex refractive index in the experiments performed in Chapter 5. This new method to measure the change in the complex refractive index induced by a dressing field will be referred to as Complex Attosecond Transient-absorption Spectroscopy (CATS).

6.2 Theory

In ATS experiments, such as those described in Chapter 5, the dynamics induced by a dressing field are imprinted upon the photoabsorption cross section and, macroscopically, the optical density (OD) of the sample [97, 132]. Measuring a change in the OD of the sample yields a rich amount of information, however it does not represent a direct measurement of all the changes induced in the sample.

To see why this is the case, consider the following scenario: a gas of atoms with density ρ that is interacting with a two-color field $\mathcal{E}(t)$ consisting of an XUV APT and an IR dressing pulse that is of moderate intensity and time delayed from the XUV APT. Moderate intensity in this particular case means that the dressing field is not strong enough to excite or ionize the ground state of the atom, but it is strong enough to further excite or ionize once the atom has been excited by the XUV APT pulse. The goal is to characterize the evolution of the electric field as it propagates through this gas medium along a direction z. Macroscopically, this can be described by Maxwell's wave equation (MWE), and in a frame that moves at the speed of light the frequency-domain MWE in cylindrical coordinates under the slowly evolving wave approximation is given by

$$\nabla_\perp^2 \tilde{\mathcal{E}}(\omega) + \frac{2i\omega}{c}\frac{\partial \tilde{\mathcal{E}}(\omega)}{\partial z} = -\frac{\omega^2}{\epsilon_0 c^2}\tilde{P}(\omega) \tag{6.1}$$

where $\tilde{\mathcal{E}}(\omega)$ is the two-color electric field in the frequency-domain ω and $\tilde{P}(\omega)$ is the polarization[14] in the frequency domain [97, 133–135]. In order to solve for $\tilde{\mathcal{E}}(\omega)$, the polarization source term $\tilde{P}(\omega)$ needs to be calculated, and this is achieved by solving the TDSE for the single atom dipole moment $d(\omega)$. This can be done in the single-active electron (SAE) approximation [136, 137], and the dipole is calculated from the time-dependent acceleration $a(t)$ given by

$$a(t) = \frac{\mathrm{d}^2\langle z\rangle}{\mathrm{d}t^2} = -\langle\psi(t)|[\hat{H},[\hat{H},\hat{z}]]|\psi(t)\rangle. \tag{6.2}$$

The dipole $\tilde{d}(\omega)$ in the frequency-domain is related to the Fourier transform $\tilde{a}(\omega)$ of $a(t)$ by the relationship $\tilde{d}(\omega) = \tilde{a}(\omega)/\omega^2$. Typically, in order to account for a finite dephasing time that naturally occurs a window function $W(t)$ is applied when Fourier transforming $a(t)$, and this results in

$$\tilde{a}(\omega) = \frac{1}{\sqrt{2\pi}}\int_{-\infty}^{\infty} a(t)W(t)e^{i\omega t}\,\mathrm{d}t. \tag{6.3}$$

From this, it is now possible to calculate the macroscopic polarization $\tilde{P}(\omega)$ as

$$\tilde{P}(\omega) = g\rho\tilde{d}(\omega) = \frac{g\rho e}{\omega^2\sqrt{2\pi}}\int_{-\infty}^{\infty} a(t)W(t)e^{i\omega t}\,\mathrm{d}t \tag{6.4}$$

where ρ is the gas density and g is the number of active electrons, which is typically 2 for rare gas atoms since there are two active electrons with the same $m = 0$ quantum number. Substituting this form of the polarization into equation 6.1 and assuming the limiting case of a linear-response the MWE becomes

$$\begin{aligned}
\frac{2i\omega}{c}\frac{\partial \tilde{\mathcal{E}}(\omega)}{\partial z} &= -\frac{\omega^2}{\epsilon_0 c^2}g\rho\tilde{d}(\omega) \\
&= -\frac{\omega^2}{\epsilon_0 c^2}\frac{g\rho\tilde{d}(\omega)}{\tilde{\mathcal{E}}(\omega)}\tilde{\mathcal{E}}(\omega).
\end{aligned} \tag{6.5}$$

---

[14]This is neglecting the polarization contribution from free electrons that have been ionized. This is reasonable because we are assuming a moderate intensity of the dressing pulse, so this term is small and can be neglected unless higher intensities are used [133].

Furthermore, in this linear-response limit, it can be assumed that $\tilde{d}/\tilde{\mathcal{E}}$ is approximately constant during propagation through the medium [97]. This allows for a simple solution given by

$$
\begin{aligned}
\tilde{\mathcal{E}}(\omega, z) &= \tilde{\mathcal{E}}_0 \exp\left(\frac{i\omega\rho}{2\epsilon_0 c}\frac{g\tilde{d}(\omega)}{\tilde{\mathcal{E}}(\omega)}z\right) \\
&= \tilde{\mathcal{E}}_0 \exp\left(\frac{i\omega\rho g}{2\epsilon_0 c}\mathrm{Re}\left[\frac{\tilde{d}(\omega)}{\tilde{\mathcal{E}}(\omega)}\right]z\right)\exp\left(-\frac{\omega\rho g}{2\epsilon_0 c}\mathrm{Im}\left[\frac{\tilde{d}(\omega)}{\tilde{\mathcal{E}}(\omega)}\right]z\right) \\
&= \tilde{\mathcal{E}}_0 \exp\left(-\tilde{\beta}(\omega)\frac{\omega}{c}z\right)\exp\left(i\tilde{n}(\omega)\frac{\omega}{c}z\right)
\end{aligned}
\tag{6.6}
$$

where the real $\tilde{n}(\omega)$ and imaginary $\tilde{\beta}(\omega)$ parts of the complex refractive index are given by

$$
\begin{aligned}
\tilde{n}(\omega) &= \frac{g\rho}{2\epsilon_0}\mathrm{Re}\left[\frac{\tilde{d}(\omega)}{\tilde{\mathcal{E}}(\omega)}\right] \\
\tilde{\beta}(\omega) &= \frac{g\rho}{2\epsilon_0}\mathrm{Im}\left[\frac{\tilde{d}(\omega)}{\tilde{\mathcal{E}}(\omega)}\right] = \frac{\rho c}{2\omega}\sigma(\omega) = \frac{c\ln 10}{2\omega L}\mathrm{OD}(\omega),
\end{aligned}
\tag{6.7}
$$

and the relationship between the imaginary part $\tilde{\beta}$, the cross section $\sigma(\omega)$, and the optical density $\mathrm{OD}(\omega)$ is also shown for a medium length L. This complex refractive index $\tilde{n} + i\tilde{\beta}$ describes the dispersion and absorption of the excited state of the gas medium in the presence of the IR dressing field. In a typical ATS experiment, only the change in absorption due to the dressing field is measured through the $\Delta\mathrm{OD}$, and this can be written in terms of the dipole as

$$
\Delta\mathrm{OD}(\omega) = \frac{g\omega\rho L}{\epsilon_0 c\ln 10}\left(\mathrm{Im}\left[\frac{\tilde{d}_{\mathrm{on}}(\omega)}{\tilde{\mathcal{E}}_{\mathrm{on}}(\omega)}\right] - \mathrm{Im}\left[\frac{\tilde{d}_{\mathrm{off}}(\omega)}{\tilde{\mathcal{E}}_{\mathrm{off}}(\omega)}\right]\right),
\tag{6.8}
$$

where the subscripts on (off) refer to when the IR dressing field is present (absent).

As can be clearly seen, only the imaginary part of the dipole is measured in a typical ATS experiment by measuring the $\Delta\mathrm{OD}$, such as was described in Chapter 5. These experiments are only able to directly measure half of the total changes induced by the dressing pulse because they are blind to the change in dispersion that arises from the change in the real part of the refractive index due to the real part of the dipole. For completeness, this change in real part of the refractive index is given by

$$
\Delta n(\omega) = \frac{g\rho}{2\epsilon_0}\left(\mathrm{Re}\left[\frac{\tilde{d}_{\mathrm{on}}(\omega)}{\tilde{\mathcal{E}}_{\mathrm{on}}(\omega)}\right] - \mathrm{Re}\left[\frac{\tilde{d}_{\mathrm{off}}(\omega)}{\tilde{\mathcal{E}}_{\mathrm{off}}(\omega)}\right]\right).
\tag{6.9}
$$

In order to directly measure the full complex refractive, and consequently both parts of the dipole, a more advanced technique is required.

140

6.2.1 Direct Measurement

To directly measure this expected change in the real part of the refractive index there are
a few methods that can be applied. In general, interferometric techniques are technically
challenging in this wavelength range because of a lack of broadband and efficient optics and
a high interferometric stability required by the relatively short wavelengths. One technique
that is often used is an interferometric method to measure the change in refractive index
as a phase shift between two arms of a Mach-Zehnder interferometer, and this concept was
introduced in detail in Chapter 4. There are a few methods to overcome these challenges,
and they generally can be divided into two different approaches: either splitting the XUV
beam after it is generated (for example, with split mirrors [54, 55]) or generating two nearly
identical XUV beams that can be interfered (for example, with a Michelson interferometer
before generation of the XUV [57]). In Chapter 3, it was shown how the latter approach can
be implemented using a $0 - \pi$ square wave phase grating (SWPG) that is able to control
the relative phase of the two generated XUV beams. Furthermore, in Chapter 4 it was
demonstrated that this setup can be used to measure the refractive index of two different
materials by measuring a phase shift between the two generated XUV sources over a broad
energy range. In that case, the phase shift induced by the sample was known *a priori*
because the ground state refractive index had been well characterized previously, and the
excellent agreement between the extracted refractive index and the known refractive index
demonstrated the validity of this technique.

In the experiment presented in Chapter 4, the complex refractive index that was measured corresponded to the ground state of a material. However, this method can be generalized to measure more than just the static ground state, and it can be used to measure a
dynamically induced change in refractive index via a dressing pulse. The basic principle is
shown schematically in figure 6.1, and it involves inducing a phase shift in only one arm of
a Mach-Zehnder interferometer that is comprised of two nearly identical XUV beams going
through the same medium. In a manner that is the same as what was presented in Chapter
4, these two XUV beams will be generated by a SWPG. The inherent interferometric stability of the SWPG due to its single-optic operation allows for measurements of phase shifts
between the two beams that correspond to a delay of only a few attoseconds (see figure
4.11 for an example) that arise due to the change in the real refractive index induced by
the dressing pulse. The relationship between the change in the real refractive index, Δn,
and the phase shift between the two beams, $\Delta\phi$, is given by

$$\Delta\phi = \frac{2\pi L \Delta n}{\lambda} \tag{6.10}$$

for a medium of length L. This phase shift between the two beams is directly measured as
a fringe shift in the spatial profile of the interference pattern in the far-field, as was shown

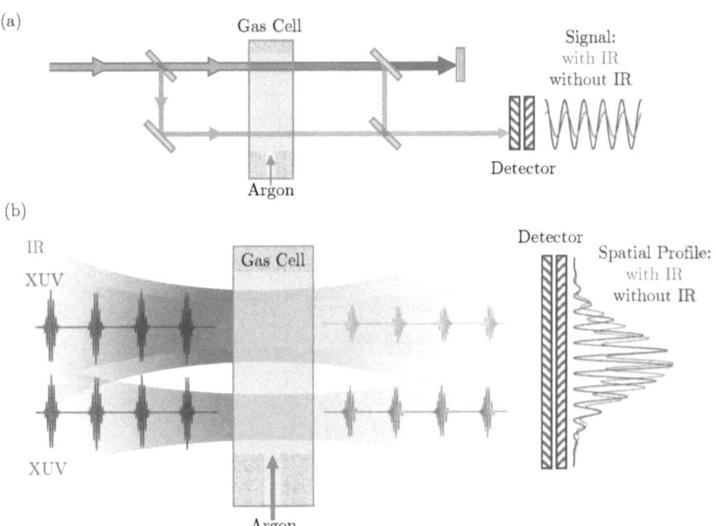

Figure 6.1: (a) Schematic of a Mach-Zehnder interferometer that is used to measure the phase shift induced by an IR dressing field introduced into one of the arms of the interferometer. (b) For the experiments described in this chapter, the two XUV sources generated by a SWPG will act as the two arms of a Mach-Zehnder interferometer, and the sample of interest will only be dressed in one of the sources by an IR field.

previously. When combined with a spectrometer this allows for the measurement of the induced phase shift as a function of photon energy, and consequently, the real part of the refractive index can be measured as a function of photon energy.

Additionally, the change in the imaginary part of the refractive index induced by the dressing pulse will lead to a change in absorption between the two beams, and this causes a change in fringe contrast in the far-field interference pattern. The relationship between the change in contrast and the change in imaginary refractive index, $\Delta\beta$, is given by

$$\Delta\beta = -\frac{c}{\omega L} \ln \left[\frac{V_0}{V} \left(1 - \sqrt{1 - \left(\frac{V}{V_0}\right)^2} \right) \right], \tag{6.11}$$

where V_0 is the fringe contrast without the dressing pulse present and V is the contrast with the dressing pulse present. This change in imaginary part of the refractive index is related to the change in absorption, ΔOD, by the relationship

$$\Delta\text{OD}(\omega) = \frac{2\omega L}{c \ln 10} \Delta\beta(\omega), \tag{6.12}$$

when the Beer-Lambert Law is assumed to hold true.

Therefore, just as in the experiment performed in Chapter 4, it is possible to directly measure the complex refractive index using a SWPG to generate an inline interferometer between two nearly identical XUV beams. The primary difference is that now we are measuring only a change in the complex refractive index induced by a dressing pulse, as opposed to the total complex refractive index which would include the ground state. This is due to the fact that this measurement is inherently differential in nature. To measure the total complex refractive index, one would have to limit the medium that is being dressed to only one of the two sources. This is technically challenging for a gas sample, so only the change in complex refractive index is measured in this chapter.

6.2.2 Indirect calculation: Kramers-Kronig Relations

Beyond a direct measurement of both parts of the complex refractive index, there is another method that can be applied when only one of the two parts of the refractive index is measured, as is the case for a normal ATS experiment. Namely, the Kramers-Kronig (KK) relations can be used to calculate one part from the other [138, 139]. These KK relations are Hilbert transforms that link the real and imaginary part of a complex function when some assumptions are met [140]. However, as will be examined later, their usefulness is limited in certain cases because they are not always directly applicable.

One of the powerful features of the KK relations is that they are fundamentally rooted in the assumption that causality holds true. This can be seen in one of the two main methods of derivation [134, 141, 142], and will be briefly shown herein. To begin, consider a dielectric medium under the influence of an electric field $\mathcal{E}(t)$. The polarization $P(t)$ is then given by

$$P(t) = \int_{-\infty}^{\infty} G(t')\mathcal{E}(t - t')\,\mathrm{d}t' \tag{6.13}$$

where $G(t')$ is the Green's function that characterizes the response of the system to a δ-function input. In the frequency domain this equation is given by

$$\tilde{P}(\omega) = \chi(\omega)\tilde{\mathcal{E}}(\omega), \tag{6.14}$$

where the susceptibility is defined in terms of the response function as

$$\chi(\omega) = \int_{-\infty}^{\infty} G(t')e^{i\omega t'}\,\mathrm{d}t'. \tag{6.15}$$

Now, invoking causality entails that the system cannot respond before the impulse. If the impulse occurs at $t = 0$, then the response must be zero for $t < 0$. Therefore, the Green's function is written as

$$G(t) = G(t)\Theta(t) \tag{6.16}$$

where $\Theta(t)$ is the Heaviside step function. Fourier transforming this relationship yields

$$\begin{aligned}
\chi(\omega) &= \chi(\omega) \circledast \left(\frac{\delta(\omega)}{2} + \frac{i}{2\pi\omega} \right) \\
&= \frac{\chi(\omega)}{2} + \frac{i}{2\pi}\mathrm{PV} \int_{-\infty}^{\infty} \frac{\chi(\omega')}{\omega - \omega'}\,\mathrm{d}\omega' \\
&= \frac{1}{i\pi}\mathrm{PV} \int_{-\infty}^{\infty} \frac{\chi(\omega')}{\omega' - \omega}\,\mathrm{d}\omega'
\end{aligned} \tag{6.17}$$

where $\circledast$ represents convolution and PV represents the Cauchy principal value. Taking the real and imaginary parts of this equation yields the KK relations for the optical susceptibility, and furthermore, they can be cast in terms of the refractive index by considering difference in output electric field betwen the two cases where the medium is present and the medium is absent. The result of this is the KK relation linking the real and imaginary part of the refractive index, and this is given by

$$\tilde{n}(\omega) - 1 = \frac{2\omega}{\pi}\mathrm{PV} \int_{0}^{\infty} \frac{\tilde{\beta}(\omega')}{\omega'^2 - \omega^2}\,\mathrm{d}\omega'. \tag{6.18}$$

Using this relationship means that only one part of the refractive index needs to be measured to fully characterize both the real and imaginary parts of the refractive index. This is a powerful tool because generally the imaginary part of the refractive index is much easier to experimentally measure. An example is shown in figure 6.2 where the real part of the refractive index is calculated from the imaginary part represented by a Lorentzian absorption line shape centered at a resonance energy ω_0. Of note in this example is the

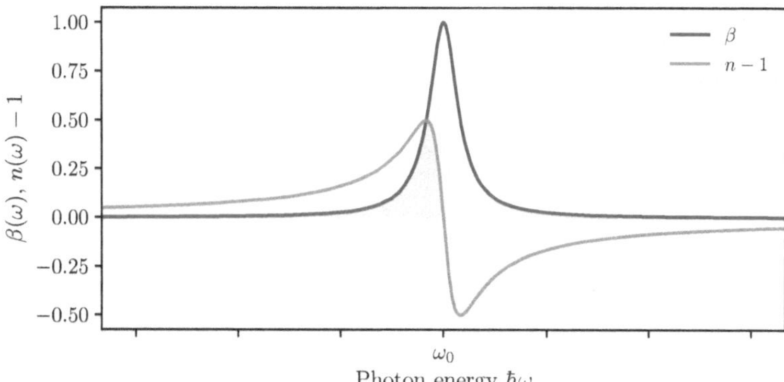

Figure 6.2: Example real $n(\omega)$ (blue) and imaginary $\beta(\omega)$ (purple) refractive index for a Lorentzian absorption line shape with a resonance at energy ω_0. The real part was calculated from the imaginary using the KK relation from equation 6.18.

difference in symmetry about the resonance energy between the two shapes, and this is a general feature of KK relations [140]. Also of note, the resolution of the KK relation depends upon the range of energies that is measured. In a similar manner to a Fourier transform, calculating one part of the refractive index with high resolution requires the other part to be measured over a large energy range. Often this means that experimental data is padded with the known refractive index outside of the measured energy range to increase resolution [72].

There is another method to derive the KK relations that is more direct, but it is less physically intuitive [131, 140, 141]. This method involves evaluating a contour integral of the susceptibility in the complex plane of ω for the contour shown in figure 6.3, and this integral is given by

$$\oint_\Gamma \frac{\chi(\omega')}{\omega' - \omega} \, d\omega' = 2\pi i \sum_k \text{Res}\left[\chi(\omega', \omega_k')\right] = 0 \tag{6.19}$$

where the residue theorem has been applied and $\text{Res}\left[\chi(\omega', \omega_k')\right]$ is the residue of χ at the pole ω_k'. Since the contour encircles no poles, the sum of residues is zero. This integral can be broken up into parts consisting of the two arcs and the line across $\text{Re}[\omega']$, and limiting cases can be applied to arrive at equation 6.17,

$$\chi(\omega) = \frac{1}{i\pi}\text{PV} \int_{-\infty}^{\infty} \frac{\chi(\omega')}{\omega' - \omega} \, d\omega'. \tag{6.20}$$

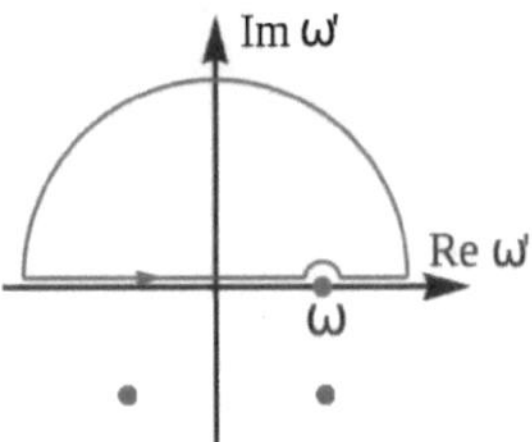

Figure 6.3: Contour Γ (green) in the complex plane of ω used to derive the KK relations. Poles are shown in red for the case of a simple harmonic oscillator.

Though the details of this derivation are not reproduced herein, the key assumption that must be met is that the susceptibility has no poles in one half plane of the complex ω plane. Mathematically, this is equivalent to the function χ being holomorphic [140]. If χ has poles in both half planes, then the sum of residues is non-zero and the KK relations do not apply. As will be shown later, this can happen in certain physical scenarios.

There were some assumptions that were made in the derivation of the KK relation in equation 6.18 that are important to scrutinize because they are not generally true. The first assumption that was made was using the linear susceptibility $\chi(\omega)$. By doing so we derived the linear KK relations, and these do not hold true when considering higher order nonlinear optical processes [140, 141]. For a higher order process, it is possible to derive KK relations that link the real and imaginary part, and these relations generally take the form given by

$$\Delta\tilde{n}(\omega,\zeta) = \frac{2\omega}{\pi}\text{PV}\int_0^\infty \frac{\Delta\tilde{\beta}(\omega',\zeta)}{\omega'^2 - \omega^2}\,\mathrm{d}\omega' \qquad (6.21)$$

where $\Delta\tilde{n}(\omega,\zeta)$ and $\Delta\tilde{\beta}(\omega,\zeta)$ are the change in the real and imaginary refractive index due to a perturbation ζ [140, 141]. For an n-th order nonlinear process, these changes in the refractive index will generally be proportional to the real and imaginary parts of the n-th order nonlinear susceptibility $\chi^{(n)}$ [11, 140, 141].

Beyond the assumption of the linear susceptibility, another scenario in which the validity of the KK relations must be examined is the case of a pump-probe experiment [140, 143, 144]. This was examined both theoretically and experimentally by Tokunaga, et. al. [143, 144], and this discussion follows from their example. Consider the change in polarization due to the pump pulse given by

$$\Delta P(t) = \chi(t) \circledast \left(\mathcal{E}(t)\Delta N(t-\tau)\right) = \int_0^\infty \chi(t')\mathcal{E}(t-t')\Delta N(t-t'-\tau)\,\mathrm{d}t' \qquad (6.22)$$

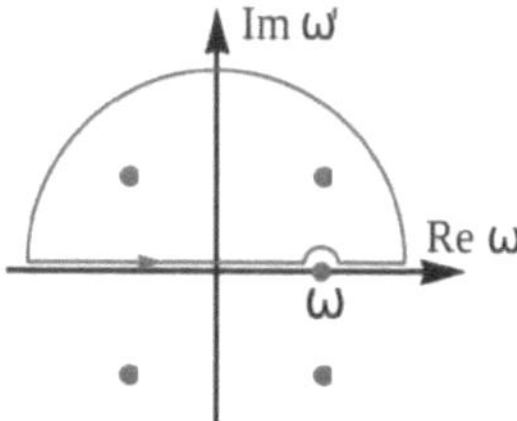

Figure 6.4: Contour (green) in the complex plane of ω used to derive the KK relations. Poles are shown in red.

where $\Delta N(t - \tau)$ represents the change in the medium due to the pump pulse for a time delay of τ. The change in susceptibility now becomes

$$\Delta\chi(\omega) = \tilde{\mathcal{E}}(\omega)\mathcal{F}[\Delta P(t)] = \tilde{\mathcal{E}}(\omega)\mathcal{F}\big[\chi(t) \circledast (\mathcal{E}(t)\Delta N(t - \tau))\big]$$
$$= \frac{\chi(\omega)}{\tilde{\mathcal{E}}(\omega)} \int_{-\infty}^{\infty} \mathcal{E}(t)\Delta N(t - \tau)e^{-i\omega t}\, dt. \tag{6.23}$$

The issue with this change in susceptibility is that within the complex plane of ω it has poles in both half planes because the integration is over all t [140, 143, 144]. This presents a problem because when we try to derive the KK relations by evaluating the contour integral over the contour shown in figure 6.4, the sum of residuals is no longer zero. The residue term is problematic because it requires knowledge of the complex function to evaluate the residue, and this means that you no longer have a simple relationship between the real and imaginary parts. Therefore, the KK relations do not hold for this case.

If some assumptions are made about $\Delta N(t-\tau)$, then the KK relations can be recovered. There are four special cases where this is possible [143, 144]:

- $\Delta N(t - \tau)$ is a constant value ΔN_0. This leads to the change in susceptibility being $\Delta\chi(\omega) = \Delta N_0\chi(\omega)$.

- $\mathcal{E}(t)$ is a δ-function $\mathcal{E}_0\delta(t)$. The change of susceptibility is then $\Delta\chi(\omega) = \Delta N(\tau)\chi(\omega)$.

- $\Delta N(t) = 0$ for $t < 0$. The change of susceptibility is then $\Delta\chi(\omega) = \frac{\chi(\omega)}{\tilde{\mathcal{E}}(\omega)} \int_0^{\infty} \mathcal{E}(t)\Delta N(t - \tau)e^{-i\omega t}\, dt$, and there are only poles in the lower half plane.

- $\Delta N(t) = 0$ for $t > 0$. The change of susceptibility is then $\Delta\chi(\omega) = \frac{\chi(\omega)}{\tilde{\mathcal{E}}(\omega)} \int_{-\infty}^0 \mathcal{E}(t)\Delta N(t - \tau)e^{-i\omega t}\, dt$, and there are only poles in the upper half plane.

Each of these special cases allows for the KK relations to be satisfied, however the last case actually means that the KK relations have the opposite sign. As a demonstration, Tokunaga et. al. experimentally tested these cases by performing a pump/probe experiment using a femtosecond frequency-domain interferometer to simultaneously measure both the real and imaginary parts of the refractive index [143]. What they found was that the KK relations were valid at most delays, however the KK relations were not valid at temporal overlap between their pump and probe pulse. This can be seen evidently in the data because the real and imaginary parts have the same symmetry, and this clearly violates the KK relations.

The breakdown of the KK relations in certain cases means that they have to be cautiously applied to pump/probe experiments, such as in ATS. There have been recent experiments that used the Fourier transform implementation of the KK relations [145] to indirectly calculate the real part from the experimentally measured change in absorption [79]. In that case, the KK relations are expected to be valid because their XUV pulse was an IAP (approximately a δ-function), and they limited their investigations to only a single time delay point that corresponded to the IR dressing field arriving after the XUV pulse. These limitations highlight the challenges that are presented by relying on an indirect calculation to reconstruct the real and imaginary parts. This is one of the reasons that a direct measurement is generally preferable because there are less assumptions that have to be made about the underlying dynamics that are of interest. To this end, the CATS technique represents a new method to directly measure both part of the refractive index change, and this will be demonstrated experimentally in the following section.

6.3 Complex Attosecond Transient-absorption Spectroscopy of Fano resonances

Now that the principle behind the CATS technique using a SWPG was established, in this section an experimental demonstration will be presented. The system of interest is the same as in Chapter 5: the $3s3p^6np$ Argon Fano resonances in the presence of an IR dressing field. This was done in an effort to examine the same physical scenario using both the traditional ATS method and the new CATS method.

6.3.1 Experimental setup

The experimental setup to demonstrate the CATS method is similar to the setup in Chapter 5, and it is shown schematically in figure 6.5. The primary differences between the two setups are the obvious presence of the SWPG to generate two XUV sources and the removal of the second harmonic generation optics that were used generate even and odd high harmonics. The dressing arm of the TABLe interferometer remains unchanged between the two experiments, and this allows for dressing of the gas sample with an intensity up to 35 TW/cm^2 at a wavelength of 1435 nm and a nominal pulse duration of 60 fs. The spot size of the dressing beam is 34 μm in diameter, and this allows for the dressing of only one of

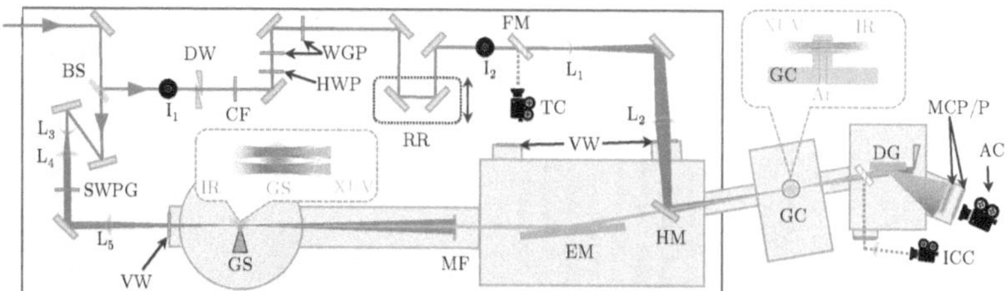

Figure 6.5: Schematic of the optical setup for the experiments described in this chapter. Insets show the two-source harmonic generation using the SWPG and the dressing of only one of the two XUV sources by an IR dressing field. **BS**: Beamsplitter, $\mathbf{I}_{1,2}$: Irises used for alignment. **DW**: Delay wedges for fine delay control. **CF**: Color filter. **HWP**: Half-wave plate. **WGP**: Wire grid polarizer. **RR**: Retro reflector for coarse delay adjustment. **FM**: Flip mirror. **TC**: Thermal camera used for alignment. $\mathbf{L}_1$: $f = -300$ mm lens. $\mathbf{L}_2$: $f = 500$ mm lens. **VW**: Vacuum window. **HM**: Hole mirror with 10 mm hole. $\mathbf{L}_3$: $f = -400$ mm lens. $\mathbf{L}_4$: $f = 500$ mm lens. $\mathbf{L}_5$: $f = 400$ mm lens. **SWPG** $0 - \pi$ square-wave phase grating. **GS**: Gas source for HHG. **MF**: Aluminum filter. **EM**: Ellipsoidal mirror. **GC**: Gas cell. **RM**: Removable mirror for *in-situ* diagnostics. **ICC**: camera for *in-situ* diagnostics. **DG**: VLS diffraction grating. **MCP/P**: Microchannel plate and phosphor. **AC**: Andor Neo 5.5 CMOS camera.

the two XUV sources that propagate through the gas cell in the target chamber.

In order to dress only one of the two XUV sources in the target chamber, as is shown in the inset of figure 6.5, the hole mirror is tilted to achieve spatial overlap with only one source. This overlap between the dressing and one of the XUV sources was verified by placing a camera at the focal plane within the target chamber. The disadvantage of this method for spatial alignment between the dressing and the XUV source is that there is now an angle introduced between the two wave fronts at the focal plane of the XUV. This angle is given by

$$\theta = \tan^{-1}\left(\frac{\Delta x}{6D}\right) = \tan^{-1}\left(\frac{\lambda f}{3dD}\right) \tag{6.24}$$

where $\Delta x = 2\lambda f/d$ is the source separation set by the SWPG of period d for a focal length f and D is the distance between the hole mirror at the focal plane of the XUV. For a SWPG with a period of $d = 2.5$ mm, a focal length of $f = 400$ mm, and a wavelength of 1435 nm, this corresponds to an angle of 0.139 mrad. This angular deviation between the two wave fronts accounts for a temporal averaging of 213 as between the XUV and the IR pulses. This averaging could be avoided in future experiments by adjusting the dressing arm focal geometry to translate the beam instead of just pointing it using a single mirror.

As mentioned previously, harmonic generation for this experiment is done at 1435 nm in

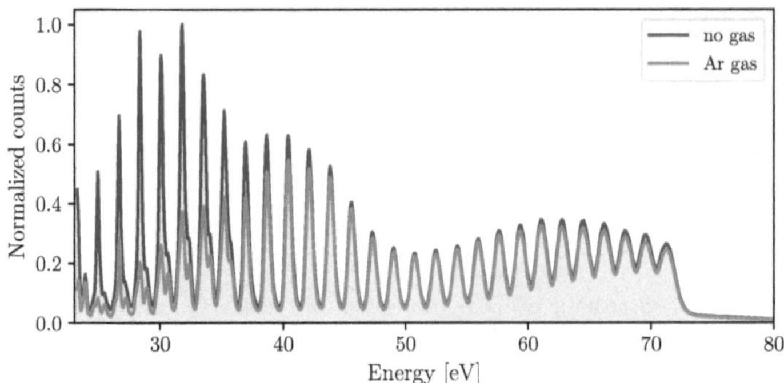

Figure 6.6: Reference harmonic spectrum used in CATS experiment with (blue) and without (purple) Ar gas present.

Ar with a SWPG to generate two sources of XUV. An example of the harmonic spectrum generated by this setup is shown in figure 6.6, and only harmonics below 72 eV are seen because a 200 nm aluminium metallic filter is used to filter out the fundamental field used for HHG. Additionally, in this figure the absorption from the sample Ar gas is shown. To do this, a gas cell is used that is 2 mm in length, 750 μm in height, and 250 μm in width. This cell was selected to allow for both XUV sources to pass through the gas medium without being clipped. For the experiments described herein, the backing pressure of the gas cell was set at 62 Torr. This interaction pressure was similar to the interaction pressure used for the experiments in Chapter 5.

To perform this experiment, it is necessary to use a SWPG to generate two XUV sources that can be interfered to act as an inline Mach-Zehnder interferometer. The details of the SWPG can be found in Chapter 3, and in this case the SWPG that was used had a grating period of $d = 2.5$mm. This entails the the source separation in the generation chamber is 459 μm for a focal length of 400 mm and a wavelength of 1435 nm. In the target chamber the source separation is reduced by a factor of 3 due to the demagnification of the ellipsoidal mirror, thus the source separation in the gas cell is only 153 μm. An example of the harmonics generated using this setup is shown in figure 6.7 (a), and the corresponding amplitude of the Fourier transform along the spatial dimension of the harmonics is shown in figure 6.7 (b). In this Fourier amplitude, clear peaks can be observed at spatial frequencies that correspond to the harmonic order of each harmonic. This depends linearly with harmonic order, as was discussed in Chapter 3. The amplitude and phase of each of these peaks will be used to define the fringe contrast and the fringe shift when comparing the

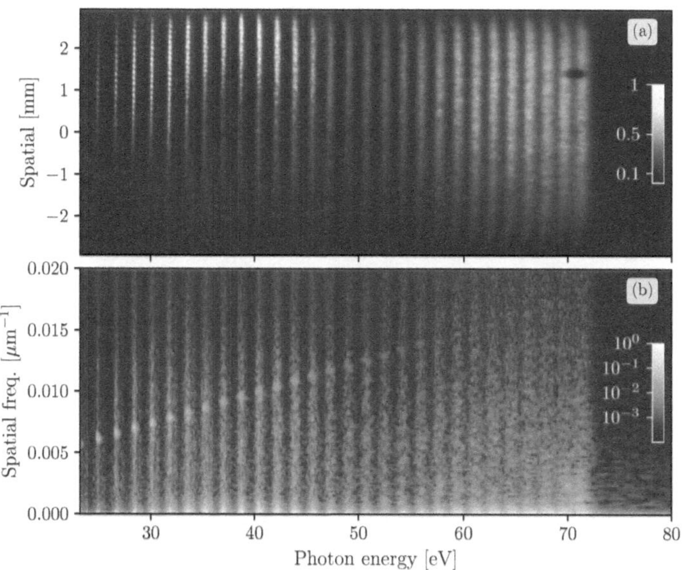

Figure 6.7: (a) Reference image of spectrometer output. Fringes along the spatial dimensions are due to interference between the two XUV sources generated by the SWPG. (b) Amplitude of Fourier transform along the spatial dimension. Peaks correspond to the spatial frequency of each harmonic order. Linear dependence of spatial frequency on photon energy is observed.

harmonic spectra with and without the dressing field present. This is done in the same manner as was done in Chapter 4 where the change in fringe contrast and shift was due the presence of a solid sample, however in this case the change in fringe contrast and shift is due the presence of a dressing field.

6.3.2 Results

Now that the theoretical background and experimental setup has been thoroughly established, attention can be turned to the experimental results. The delay dependent change in the real and imaginary parts of the refractive index will be shown for both moderate and high dressing field intensities, and the extracted dipole amplitude and phase will be shown. Additionally, the applicability, or lack thereof, of the KK relations will be examined.

To begin, the delay dependent results are shown in figure 6.8 for dressing intensities of $12\ \mathrm{TW/cm^2}$ and $33\ \mathrm{TW/cm^2}$. In figures 6.8 (a) and (b), the ΔOD is calculated in a similar manner as was done in Chapter 5 and is comparable to the results that were presented previously. The primary difference between this calculation and what was shown previously is the fact that only one of the two sources being dressed needs to be accounted for to correctly extract the amplitude of the ΔOD signal. If the ratio of the total power in each source is $R = I_1/I_2$, then the true change in transmission ΔT induced by the dressing field can be written in terms of the measured change in total transmission $\Delta \tilde{T}$ as

$$\Delta \tilde{T} = \frac{I_1 + \Delta T I_2}{I_1 + I_2} = \frac{R + \Delta T}{1 + R}. \tag{6.25}$$

Using this relationship the true ΔOD(ω) is given by

$$\begin{aligned}
\Delta \mathrm{OD}(\omega) &= -\log(\Delta T(\omega)) = -\log((1 + R)\Delta \tilde{T}(\omega) - R) \\
&= -\log\left((1 + R)\frac{I_{\mathrm{on}}(\omega)}{I_{\mathrm{off}}(\omega)} - R\right)
\end{aligned} \tag{6.26}$$

where I_{on} and I_{off} is the harmonic amplitude with and without the dressing field, respectively. This ratio R is a free parameter that is used to account for the difference in amplitude between the two sources, and it is generally found to be around $R = 0.8$. To calculate the harmonic amplitude $I_{\mathrm{on,off}}(\omega)$, the spatial profile of the harmonics is integrated over to yield the amplitude as a function of energy.

As can be seen in the ΔOD in figures 6.8 (a) and (b), there are two main features that can observed at 26.6 eV and 28.5 eV. These features correspond to the $3s3p^6np$ autoionizing (Fano) resonances for $n = 4$ at 26.6 eV and $n = 6$ at 28.5 eV, which have been extensively explained in section 5.2. For this particular experiment, only these two resonances are observed because the harmonic comb that was used only consisted of odd harmonics, as opposed to a comb of even and odd harmonics that can act as a pseudo-continuum of energies if each harmonic has sufficient bandwidth, see section 5.5.1. Thus, only features that are present within the bandwidth of each harmonic can be observed in this experiment. Consequently, the wavelength of 1435 nm was selected to place two resonances within the bandwidth of neighboring harmonics.

The general interpretation of these features in the ΔOD was explored in terms of the dipole control model (DCM) and Floquet theory in sections 5.3 and 5.4. Generally speaking, the DCM explains the attenuation and phase shift of the dipole in the time domain due to the influence of the dressing field, and the Floquet theory explains the presence of light-induced states (LIS) that appear in the ΔOD. For the dressing intensities used in this experiment, the dipole is strongly suppressed by the dressing field, as was shown in the intensity dependence of the ΔOD in figures 5.20 and 5.23. The Floquet theory accounts for the presence of LIS in the absorption spectrum that are generally observed between the bright np resonances. In this case, since we can only observe changes within the bandwidth of each harmonic, the same LISs are not observed. However, Floquet theory should still be able to account for the shifting of levels in the dressed atom picture. The observed delay dependent ΔOD is consistent with what was observed previously, see section 5.5.4, and the primary difference being that only features near the bright $4p$ and $6p$ states are observed.

Moving beyond the typical ATS measurement, in figures 6.8 (c) and (d) the change in fringe contrast is used to extract the ΔOD. This is done by Fourier transforming the spatial structure of the harmonics and taking the amplitude of the spatial frequency corresponding to the harmonic energy. From this, the fringe contrast V is defined as

$$V = \frac{I_{\max} - I_{\min}}{I_{\max} + I_{\min}} = \frac{I_{\mathrm{amp}}}{I_{\mathrm{mean}}}. \tag{6.27}$$

If V (V_0) is the contrast with (without) the dressing field, then the imaginary part $\Delta\beta$ can be calculated using equation 6.11. From this, the ΔOD can then be calculated using equation 6.12, and this is exactly what is shown in figures 6.8 (c) and (d) for a dressing intensity of 12 TW/cm^2 and 33 TW/cm^2, respectively. Comparing the ΔOD extracted using this method and the typical method, it can be seen that there is overall good agreement between the two methods. The primary difference is that extracting the ΔOD using the fringe contrast can only be done where the fringe visibility is sufficiently high within the bandwidth of each harmonic. This limits the energy range that can be used to extract the ΔOD, however it still represents a significant fraction of the harmonic bandwidth.

The more interesting component of this measurement is shown in figures 6.8 (e) and (f), which are the phase shift $\Delta\phi$ as a function of energy and delay that is extracted from the fringe shift of the spatial interference pattern. This is done a in similar manner as the ΔOD by Fourier transforming the spatial structure of the harmonics and taking the phase of the spatial frequency corresponding to that harmonic energy. The difference in this phase

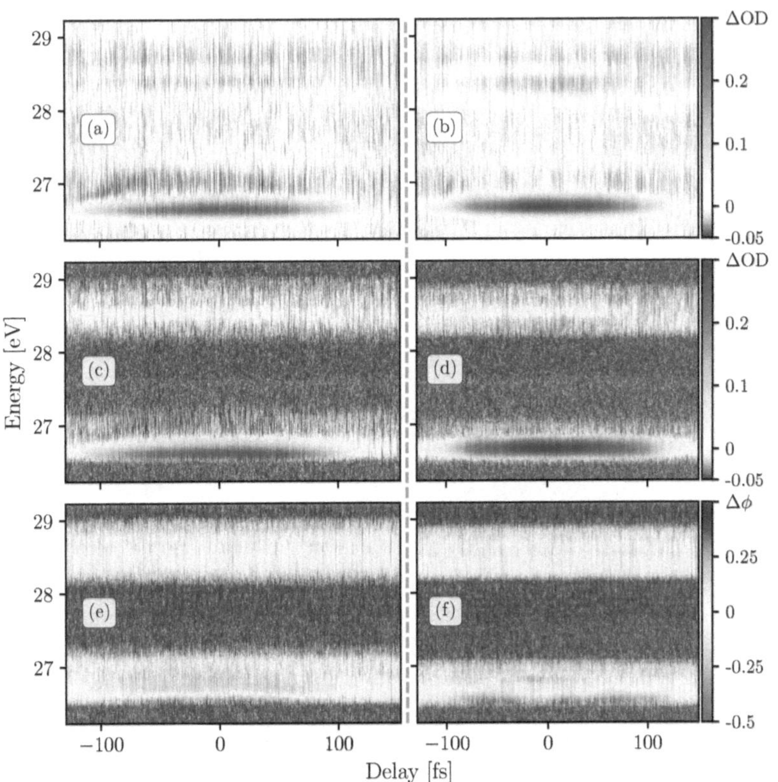

Figure 6.8: (a),(b) ΔOD calculated when ignoring spatial structure of harmonics. (c),(d) ΔOD calculated using the change in fringe contrast induced by the dressing field. (e),(f) Phase shift $\Delta\phi \propto \Delta n$ calculated from the fringe shift induced by the dressing field. Figures on left correspond to a dressing intensity of 12 TW/cm^2, and figures on the right correspond to a dressing intensity of 33 TW/cm^2.

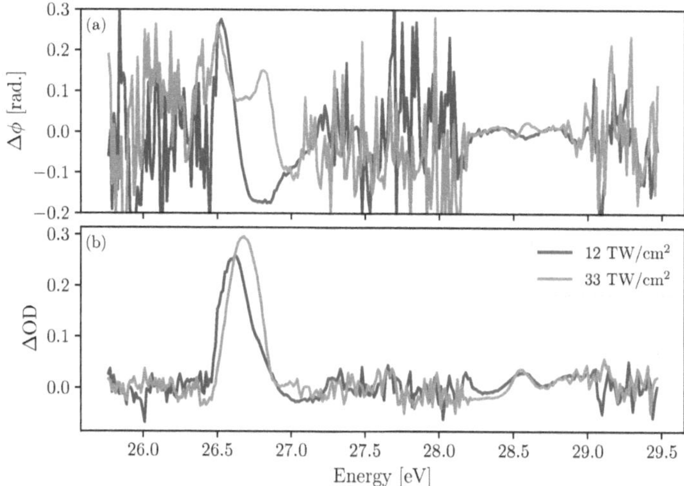

Figure 6.9: (a) Line outs of the dressing field induced phase $\Delta\phi$ for the two intensities 12 TW/cm^2 (purple) and 33 TW/cm^2 (blue). (b) Line outs of the ΔOD measured using the dressing field induced change in fringe contrast for the two intensities.

with and without the dressing field gives the dressing field induced phase shift $\Delta\phi$ that is proportional to the real part of the refractive index Δn, see equation 6.10. As can be seen, there is a clear delay dependent phase shift that confirms the feasibility of measuring both parts of the refractive index using CATS.

Additionally, in the delay dependent phase shift $\Delta\phi$ there is a clear difference in the induced phase between the two intensities. To examine this further, it is useful to look at line outs along the energy axis for a delay corresponding to $\tau = 0$, and this is shown in figure 6.9. At the lower intensity, the phase shift in the vicinity of the $4p$ resonance has an asymmetric dispersive line shape. However, at the higher intensity the structure of the phase difference is more complicated. On the low energy side near 26.5 eV there is agreement between the two intensities, however a second peak appears at 26.8 eV. This second peak can attributed to a LIS based upon its energy because it is separated by approximately $2\omega = 1.74$ eV from the higher $6p$ resonance at 28.5 eV. The fact that this transition would be three photons (1 XUV - 2 IR) from the ground state means that it would depend strongly on the dressing intensity [146], and thus it would only be expected to appear at the higher dressing intensity near temporal overlap between the XUV and IR pulses. This is exactly what is observed in

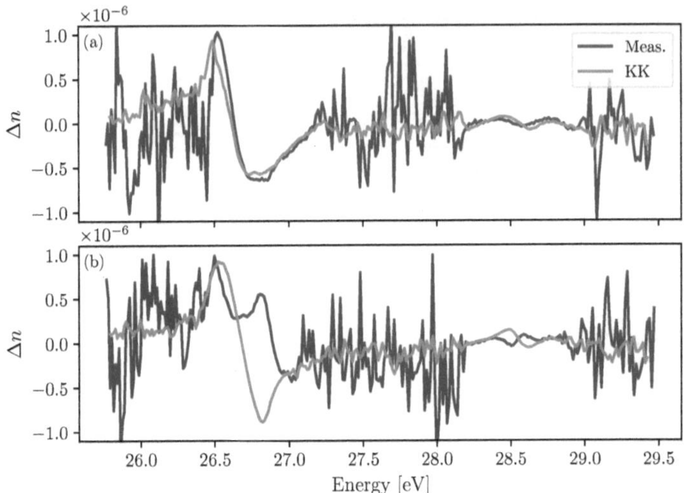

Figure 6.10: Indirect calculation of Δn using the nonlinear KK relation for (a) the lower intensity of 12 TW/cm^2 and (b) the higher intensity 33 TW/cm^2.

figure 6.8. At higher energies near the $6p$ resonance, there is also a clear difference in the induced phase between the two intensities. At lower intensities there is a broad decrease in phase difference, however at higher intensities this becomes more of a dispersive line shape and changes sign.

In terms of the ΔOD line outs shown in figure 6.9 (b), there is also a clear difference between the two intensities that were measured. Near the $4p$ resonance the line shape for both intensities is qualitatively similar because there appears to be only a single peak for both intensities, however the peak position and shape is different between the two cases. The higher intensity case shows the peak shifted to higher energies, and this can most likely be attributed to the presence of an unresolved peak in ΔOD due to the LIS that was observed in the induced phase difference. Near the $6p$ resonance there is less of a difference between the two intensities. Generally speaking, they have the same shape, however the higher intensity case shows a slightly narrower peak.

Now that the phase difference $\Delta\phi$ and the change in absorption ΔOD has been characterized, one question to ask is: can this same result be achieved using only the KK relations? To explore this option, the nonlinear KK relation given by equation 6.21 is used to indirectly calculate the change in real refractive index $\Delta n = \lambda\Delta\phi/2\pi L$, and this is shown in

figure 6.10. For the lower intensity the answer appears to be yes, the KK relations can be
applied in this case to indirectly calculate the real part from the imaginary. The primary
difference between the calculation and the experimental values is near the $6p$ resonance
where the shape is reproduced, but the peak positions are slightly shifted. This is most
likely due to the small signal in this region that is close to the noise floor. That being said,
the story is quite different at higher intensities. In this case, the KK relation completely
fails to reproduce the experimental data. The peak near 26.5 eV is reasonably reproduced,
however the LIS peak at 26.8 eV cannot be reproduced by the KK relation. Additionally,
the features seen near the $6p$ resonance at 28.5 eV have the opposite sign. The failure of
the KK relations is not surprising in this case because they fundamentally connect shapes
of opposite symmetry, and the fact the ΔOD was symmetric for both intensities means that
it can only produce an asymmetric line shape for the real part. This break down of the
KK relations near overlap has been experimentally observed before at visible wavelengths
[143]. However, the intensity dependent breakdown of the KK relation will require more
theoretical effort to fully understand.

A final feature that can be extracted using CATS is the amplitude and phase of the
dipole from these measurements. In order to extract the dipole, knowledge of the electric
field $\tilde{\mathcal{E}}(\omega)$ spectral amplitude and phase needs to be known. This can be seen in the
relationships given by equations 6.9 and 6.8. The amplitude is directly measured in the
experiment, however the phase in unknown. In principle, the phase can be measured using
RABBITT [144], or it could be calculated [147]. In lieu of that, the change in polarizability
can be extracted, where the polarizability is given by $\alpha = \tilde{d}/\tilde{\mathcal{E}}$. Thus, the amplitude of $\Delta\alpha$
is given by

$$|\Delta\alpha|^2 = \left(\frac{2\epsilon_0}{g\rho}\Delta n\right)^2 + \left(\frac{\epsilon_0 c \ln 10}{g\rho\omega L}\Delta\text{OD}\right)^2,\tag{6.28}$$

and the phase of $\Delta\alpha$ is given by

$$\Phi_{\Delta\alpha} = \arg\left[\frac{2\epsilon_0}{g\rho}\Delta n + i\frac{\epsilon_0 c \ln 10}{g\rho\omega L}\Delta\text{OD}\right].\tag{6.29}$$

The change in amplitude and phase of $\Delta\alpha$ is shown in figure 6.11 for the two intensities.
To highlight the differences between the two intensities, line outs of $|\Delta\alpha|$ and $\Phi_{\Delta\alpha}$ near
temporal overlap are shown in figure 6.12.

6.4 Conclusion

In this chapter, the change in real and imaginary refractive index was expounded theoretically to motivate the need for a new method to measure the complex refractive index in
a transient absorption experiment. The method developed to do this is Complex Attosec-

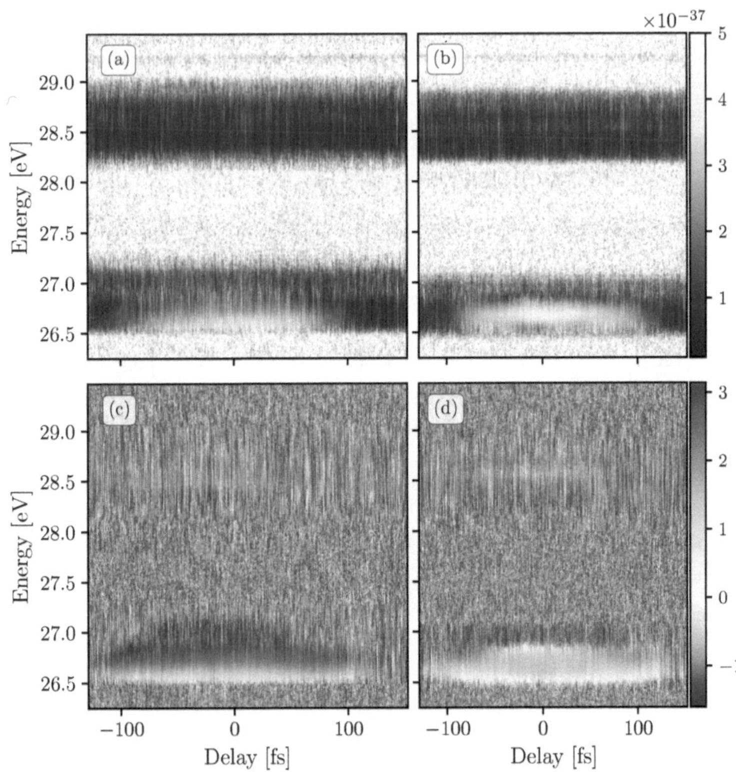

Figure 6.11: (a),(b) Amplitude of change in polarizability $\Delta\alpha$ in units of Fcm^2. For context, the atomic unit of polarizability is 1.648×10^{-37} Fcm^2. (c),(d) Phase of change in polarizability $\Delta\alpha$ in units of rad. Figures on left correspond to a dressing intensity of 12 TW/cm^2, and figures on right correspond to a dressing intensity of 33 TW/cm^2.

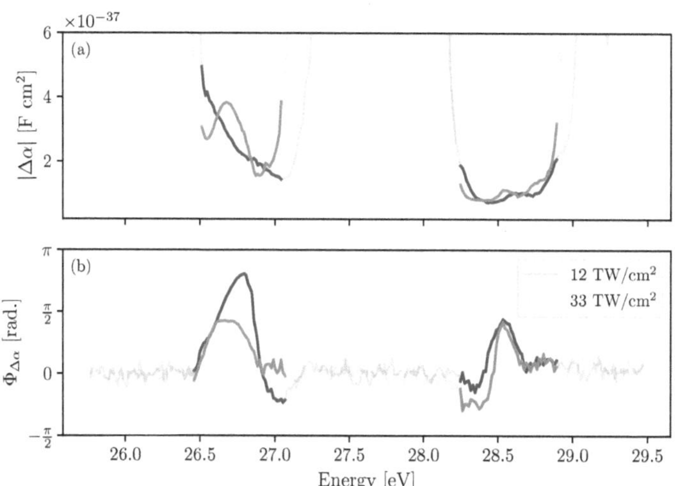

Figure 6.12: (a) Line outs of $|\Delta\alpha|$ at temporal overlap, $\tau = 0$, for dressing intensities of 12 TW/cm^2 (purple) and 33 TW/cm^2 (blue). (b) Line out of the phase $\Phi_{\Delta\alpha}$ for the two intensities. The greyed out lines indicate where the harmonic amplitude was zero, and therefore the extracted values are not meaningful.

ond Transient-absorption Spectroscopy (CATS). CATS leverages a 0-π SWPG to generate a reference XUV field that can be used to directly measure both parts of the refractive index. This method was verified experimentally by studying the argon Fano resonances in the presence of a IR dressing field of moderate and high intensity. The demonstration of CATS allows for the direct measurement of the complex refractive index in a host experiments, and paves the way for future research by measuring a feature that was previously only indirectly accessible in typical ATS experiments.

Chapter 7
CONCLUSION

The work presented herein represents the culmination of many years worth of research, and it sets the foundation for future experiments. In particular, demonstrating the ability to directly measure the full complex refractive index of ground states and excited states opens the door to a new class of experiments that were not previously possible. Each Chapter builds towards this crescendo, starting with Chapter 2 where the TABLe was introduced. This beamline was designed to have the unique capabilities necessary to perform the experiments that would follow. Starting with the experiments done in Chapter 3, which demonstrated the use of a 0-π SWPG to generate two XUV sources and control the relative phase between them. This interferometric control is critical to the retrieval of the complex refractive index that was done in Chapter 4. In this Chapter, it was demonstrated that the SWPG can be used to measure the ground state refractive index over a broad energy range of two different materials. This measurement of the ground state refractive index naturally begs the question: Can the excited state be measured too? To study this, a suitable physical system was selected for study, and that was the argon Fano resonances. An initial experiment was performed using standard transient absorption to fully characterize this system in our experimental conditions, and this was detailed in Chapter 5. Following this, the methods developed in Chapter 4 were extended to two-source transient absorption, and this was termed the CATS method. CATS was used to measure the excited state dynamics of the laser dressed argon Fano resonances in Chapter 6. The success of CATS in this study is what conclusively demonstrates the capability of CATS to be a viable method for future experiments to gain access to both the real and imaginary parts of the refractive index. CATS was demonstrated in a gas phase experiment, however it can easily be applied to condensed matter experiments in either a transmission or a reflection geometry. This allows for full and direct characterization of the induced dynamics, and this can be a powerful tool to further advance attosecond physics.